T0313146

Advanced Spatial Modeling with Stochastic Partial Differential Equations Using R and INLA

Elias Krainski
Virgilio Gómez-Rubio
Haakon Bakka
Amanda Lenzi
Daniela Castro-Camilo
Daniel Simpson
Finn Lindgren
Håvard Rue

CRC Press
Taylor & Francis Group
Boca Raton London New York

CRC Press is an imprint of the
Taylor & Francis Group, an **informa** business

CRC Press
Taylor & Francis Group
6000 Broken Sound Parkway NW, Suite 300
Boca Raton, FL 33487-2742

© 2019 by Taylor & Francis Group, LLC
CRC Press is an imprint of Taylor & Francis Group, an Informa business

No claim to original U.S. Government works

Printed on acid-free paper
Version Date: 20181115

International Standard Book Number-13: 978-1-138-36985-6 (Hardback)

Visit the Taylor & Francis Web site at
http://www.taylorandfrancis.com

and the CRC Press Web site at
http://www.crcpress.com

This book is dedicated to all the patient users of the SPDE approach and `R-INLA` in its early stage. The development really benefited from all your feedback. Thank you!

Contents

Preface

This book grew out of a tutorial written by Elias T. Krainski, which he started in 2013 together with his PhD-studies at NTNU, Trondheim, Norway. The tutorial has since then been expanded continuously, based on response from the many users and based on new developments.

Lindgren et al. (2011) describe an approximation to continuous spatial models with a Matérn covariance that is based on the solution to a stochastic partial differential equation (SPDE). This approximation is computed using a sparse representation that can be effectively implemented using the integrated nested Laplace approximation (INLA, Rue et al., 2009).

This book will show you how to fit models that contain at least one effect specified with an SPDE using the INLA package for the R software for statistical computing. An SPDE based model will be used to define random effects over continuous domains in one or two dimensions. The usual application is data whose geographical location is explicitly considered in the analysis.

This book explores INLA functionalities through examples, and it is structured as follows. Chapter 1 provides an introduction to the integrated nested Laplace approximation and its associate INLA package for the R programming language. Chapter 2 introduces Gaussian random fields and the SPDE framework, develops an example on a toy dataset and works through some examples of building a mesh. Here, an example with non-Gaussian data is also discussed. Then three examples on the use of models with several likelihoods are developed in Chapter 3. These include a measurement error model, a coregionalization model and considering part or the entire linear predictor from one outcome in a linear predictor of another one. Point pattern analysis is included in Chapter 4 using a log-Gaussian Cox process. Non-stationary spatial models are developed in Chapter 5, which includes inclusion of covariates in the covariance parameters and barrier models. Chapter 6 focuses on survival analysis, and models for extremes and non-standard likelihoods are discussed here. Space-time models are described in detail in Chapter 7. Some applications of space-time models are developed in Chapter 8. Two appendices are included at the end with a summary of the notation used in the book and information about the R packages required to reproduce the examples in the book.

The introduction in Chapter 1 can be used as a starting point for the integrated nested Laplace approximation and the INLA package. Chapter 2 tries to explain some of the theoretical details behind the SPDE approach by developing two examples. Going through the more theoretical details may require some background on stochastic processes, but the applications of the SPDE approach are described in detail in the examples in this chapter and throughout the book.

This book focuses on SPDE models with INLA but it does not cover the basics of Bayesian inference or spatial analysis. For this, Bivand et al. (2013) give a thorough description of spatial analysis in R. Banerjee et al. (2014) cover Bayesian inference for different types of spatial models in detail. Blangiardo and Cameletti (2015) and Zuur et al. (2017) give an introduction to INLA and discuss spatial and spatio-temporal models. Wang et al. (2018) and Gómez-Rubio (2019) provide a good introduction to INLA and modeling with the INLA package, which are a good resource to learn about INLA.

There are some other resources available on-line or in the INLA package. Lindgren and Rue (2015) is an excellent tutorial available at http://www.r-inla.org/examples/tutorials/spde-tutorial-from-jss. If you are in a rush to fit a simple geostatistical model, please see the vignette available in the INLA package which can be loaded by typing vignette(SPDEhowto) or a one dimensional example by typing vignette(SPDE1d). A mesh building demonstration Shiny app can be opened by typing meshbuilder().

Finally, a Gitbook version of this book is available from the book website at http://www.r-inla.org/spde-book. Here, R code and datasets used in the examples and figures of this book are also available. We have tried to use color-blind friendly palettes throughout the book using packages RColorBrewer and viridisLite, but this can be easily changed in the provided R code.

████

Acknowledgement

We would like to thank Sarah Gallup and Helen Sofaer for some English review in the tutorial that originated this book. Our thanks to several people who brought nice problems and questions to the INLA discussion forum at http://www.r-inla.org/comments-1 and directly to us. Finally, we are grateful to John Kimmel and CRC for being supportive about the publication of this book and for his help throughout the publication process.

Elias T. Krainski was supported by a grant from the Norwegian Research Council, during the years 2013-2016. Virgilio Gómez-Rubio has been partly supported by grant SBPLY/17/180501/000491, awarded by Consejería de Educación, Cultura y Deportes (JCCM, Spain) and FEDER, grant MTM2016-77501-P, awarded by Ministerio de Economía y Competitividad (Spain) and a grant to support research groups from Universidad de Castilla-La Mancha (Spain).

This book has been written using the `bookdown` package and `R` markdown. Map data from Albacete copyrighted OpenStreetMap contributors and is available from `https://www.openstreetmap.org`.

What this book is and isn't

This book has a somewhat unusual format, and we need to clarify this to avoid confusing readers. The goal of this book is to provide several tools for cutting-edge research in applied spatial modeling, in under 300 pages. To achieve this goal we made a few unusual decisions when writing this book:

- No detailed introduction: there is no detailed introduction to spatial modeling or spatial processes in this book. We assume the reader is familiar with this literature or has other sources to depend on.

- The explanations of the models are limited to the minimum, in order to include more code examples in the book.

- No detailed applied discussions: each chapter contains only a simple interpretation of the results.

- The first two chapters should be read (at least superficially) before any of the other chapters, but otherwise, the chapters can be read in any order.

The book is centered around R code, with the goal of being "useful" to applied users:

- All the code should run quickly, even for complex multi-likelihood space-time problems.

- Additional reading is required to understand the statistical properties of the models, but the reader can use and expand on the code to study these statistical properties.

- The codes in the different chapters are independent, allowing the user to copy the code from only one chapter.

This is advanced applied statistical modeling, hence some of the models can be difficult to work with. We will continue to develop this book based on user feedback, through the online repository http://www.r-inla.org/spde-book.

1

The Integrated Nested Laplace Approximation and the `R-INLA` package

1.1 Introduction

In this introductory chapter a brief summary of the integrated nested Laplace approximation (INLA, Rue et al., 2009) is provided. The aim of this chapter is to introduce the INLA methodology and the main features of the associated `INLA` package (also called `R-INLA`) for the `R` programming language. Different models will be fitted to a simulated dataset in order to show the main steps to fit a model with the INLA methodology and the `INLA` package.

This introduction to INLA and the `INLA` package covers the basics as well as some advanced features. The aim is to provide the reader with a general overview that will be useful to follow the other chapters in the book on spatial models with the SPDE methodology. Other recent works that describe INLA are the following: Blangiardo and Cameletti (2015) provide an introduction to the main INLA theory and an extensive description of many spatial and spatio-temporal models. Wang et al. (2018) provide a detailed description of INLA with a focus on general regression models. Similarly, Gómez-Rubio (2019) describes the underlying INLA methodology and describes many different models and computational aspects.

1.2 The INLA method

Rue et al. (2009) develop the integrated nested Laplace approximation (INLA) for approximate Bayesian inference as an alternative to traditional Markov chain Monte Carlo (MCMC, Gilks et al., 1996) methods. INLA focuses on models that can be expressed as latent Gaussian Markov random fields (GMRF) because of their computational properties (see Rue and Held, 2005, for details). Not surprisingly, this covers a wide range of models and recent reviews of INLA and its applications can be found in Rue et al. (2017) and Bakka et al. (2018).

The INLA framework can be described as follows. First of all, $\mathbf{y} = (y_1, \ldots, y_n)$ is a vector of observed variables whose distribution is in the exponential family (in most cases), and the mean μ_i (for observation y_i) is conveniently linked to the linear predictor η_i using an appropriate link function (it is also possible to link the predictor to e.g. a quantile). The linear predictor can include terms on covariates (i.e., fixed effects) and different types of random effects. The vector of all latent effects will be denoted by $\mathbf{x}$, and it will include the linear predictor, coefficients of the covariates, etc. In addition, the distribution of $\mathbf{y}$ will likely depend on some vector of hyperparameters $\boldsymbol{\theta}_1$.

The distribution of the vector of latent effects $\mathbf{x}$ is assumed to be Gaussian Markov random field (GMRF). This GMRF will have a zero mean and precision matrix $\mathbf{Q}(\boldsymbol{\theta}_2)$, with $\boldsymbol{\theta}_2$ a vector of hyperparameters. The vector of all hyperparameters in the model will be denoted by $\boldsymbol{\theta} = (\boldsymbol{\theta}_1, \boldsymbol{\theta}_2)$.

Furthermore, observations are assumed to be independent given the vector of latent effects and the hyperparameters. This means that the likelihood can be written as

$$\pi(\mathbf{y}|\mathbf{x}, \boldsymbol{\theta}) = \prod_{i \in \mathcal{I}} \pi(y_i|\eta_i, \boldsymbol{\theta}).$$

Here, η_i is the latent linear predictor (which is part of the vector $\mathbf{x}$ of latent effects) and set $\mathcal{I}$ contains indices for all observed values of $\mathbf{y}$. Some of the values may not have been observed.

The aim of the INLA methodology is to approximate the posterior marginals of the model effects and hyperparameters. This is achieved by exploiting the computational properties of GMRF and the Laplace approximation for multidimensional integration.

The joint posterior distribution of the effects and hyperparameters can be expressed as:

$$\pi(\mathbf{x}, \boldsymbol{\theta}|\mathbf{y}) \propto \pi(\boldsymbol{\theta})\pi(\mathbf{x}|\boldsymbol{\theta}) \prod_{i \in \mathcal{I}} \pi(y_i|x_i, \boldsymbol{\theta}) \tag{1.1}$$

$$\propto \pi(\boldsymbol{\theta})|\mathbf{Q}(\boldsymbol{\theta})|^{1/2} \exp\{-\frac{1}{2}\mathbf{x}^\top \mathbf{Q}(\boldsymbol{\theta})\mathbf{x} + \sum_{i \in \mathcal{I}} \log(\pi(y_i|x_i, \boldsymbol{\theta}))\}.$$

Notation has been simplified by using $\mathbf{Q}(\boldsymbol{\theta})$ to represent the precision matrix of the latent effects. Also, $|\mathbf{Q}(\boldsymbol{\theta})|$ denotes the determinant of that precision matrix. Furthermore, $x_i = \eta_i$ when $i \in \mathcal{I}$.

The computation of the marginal distributions for the latent effects and hyperparameters can be done considering that

$$\pi(x_i|\mathbf{y}) = \int \pi(x_i|\boldsymbol{\theta}, \mathbf{y})\pi(\boldsymbol{\theta}|\mathbf{y})d\boldsymbol{\theta},$$

and

$$\pi(\theta_j|\mathbf{y}) = \int \pi(\boldsymbol{\theta}|\mathbf{y})d\boldsymbol{\theta}_{-j}.$$

Note how in both expressions integration is done over the space of the hyperparameters and that a good approximation to the joint posterior distribution of the hyperparameters is required. Rue et al. (2009) approximate $\pi(\boldsymbol{\theta}|\mathbf{y})$, denoted by $\tilde{\pi}(\boldsymbol{\theta}|\mathbf{y})$, and use this to approximate the posterior marginal of the latent parameter x_i as:

$$\tilde{\pi}(x_i|\mathbf{y}) = \sum_k \tilde{\pi}(x_i|\boldsymbol{\theta}_k, \mathbf{y}) \times \tilde{\pi}(\boldsymbol{\theta}_k|\mathbf{y}) \times \Delta_k.$$

Here, Δ_k are the weights associated with a vector of values $\boldsymbol{\theta}_k$ of the hyperparameters in a grid.

The approximation $\tilde{\pi}(\boldsymbol{\theta}_k|\mathbf{y})$ can take different forms and be computed in different ways. Rue et al. (2009) also discuss how this approximation should be in order to reduce the numerical error.

1.2.1 The R-INLA package

The INLA methodology is implemented in the INLA package (also known as R-INLA package), whose download instructions are available from the main INLA website at http://www.r-inla.org. INLA is available as an R package for Windows, Mac OS and Linux from its own repository, as it is not available on CRAN yet. The testing version can be downloaded as:

```
# Set CRAN mirror and INLA repository
options(repos = c(getOption("repos"),
  INLA = "https://inla.r-inla-download.org/R/testing"))
# Install INLA and dependencies
install.packages("INLA", dependencies = TRUE)
```

The stable version can be downloaded by replacing testing by stable in the code above. This book is compiled with the testing version, which may include (newer) features required in some parts of this book.

The main function in the INLA package is inla(), which provides a simple method of model fitting. This function works in a similar way as the glm() or gam() functions. A formula is used to define the model, with fixed and random effects, together with a data.frame. In addition, both generic options on how to compute the results and specific model settings can be passed into the call. A simple example is provide in Section 1.3.

1.3   A simple example

Here, we develop a simple example to illustrate the INLA methodology. We will use the INLA package through the SPDEtoy dataset. This dataset is extensively analyzed in Section 2.8 using spatial models, but we will focus on simpler regression models now. The SPDEtoy dataset contains simulated data from a continuous spatial process in the unit square. This would mimic, for example, typical spatial data such as temperature or rainfall that occur continuously over space but that are only measured at particular locations (usually, where stations are placed). Table 1.1 includes a summary of the three variables included in the dataset.

TABLE 1.1: Variables in the SPDEtoy dataset.

Variable	Description
y	Simulated observations at the locations.
s1	x-coordinate in the unit square.
s2	y-coordinate in the unit square.

The dataset can be loaded as follows:

```
library(INLA)
data(SPDEtoy)
```

The following code will take the original data, create a SpatialPointsDataFrame (Bivand et al., 2013) to represent the locations and create a bubble plot on variable y:

```
SPDEtoy.sp <- SPDEtoy
coordinates(SPDEtoy.sp) <- ~ s1 + s2

bubble(SPDEtoy.sp, "y", key.entries = c(5, 7.5, 10, 12.5, 15),
       maxsize = 2, xlab = "s1", ylab = "s2")
```

Figure 1.1 displays the values of the observations using a bubble plot. Here, a clear trend in the data can be observed, with more values observed close to the bottom left corner.

The first model we fit to the SPDEtoy dataset with INLA is a linear regression on the coordinates. This will be done using function inla(), which takes similar arguments to the lm(), glm() and gam() functions (to mention a few).

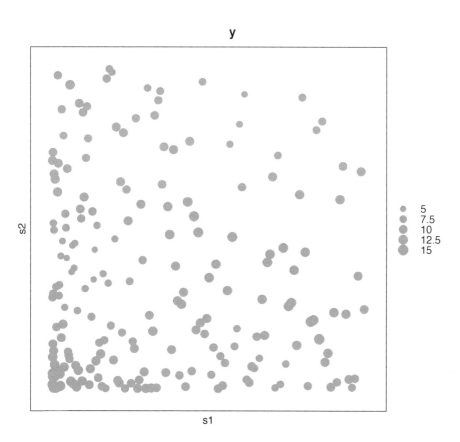

FIGURE 1.1 Bubble plot of the SPDEtoy dataset.

Under this model, observation y_i at location $\mathbf{s}_i = (s_{1i}, s_{2i})$ is assumed to be distributed as a Gaussian variable with mean μ_i and precision τ. The mean μ_i is assumed to be equal to $\alpha + \beta_1 s_{1i} + \beta_2 s_{2i}$, with α the model intercept and β_1 and β_2 the coefficients of the covariates. By default, the prior on the intercept is a uniform distribution; the prior on the coefficients is also a Gaussian with zero mean and precision 0.001, and the prior on the precision τ is a Gamma with parameters 1 and 0.00005. We can change the priors as discussed in Section 1.4.2.

We note that this example is used here simply to illustrate how to compute results with INLA. Furthermore, most of the models in this section are not reasonable spatial models since they lack rotation invariance of the coordinate system, which is a desirable property in many spatial models (see Chapter 2 for details).

The model can be stated as follows;

$$
\begin{aligned}
y_i &\sim N(\mu_i, \tau^{-1}), \ i = 1, \ldots, 200 \\
\mu_i &= \alpha + \beta_1 s_{1i} + \beta_2 s_{2i} \\
\alpha &\sim \text{Uniform} \\
\beta_j &\sim N(0, 0.001^{-1}), \ j = 1, 2 \\
\tau &\sim Ga(1, 0.00005).
\end{aligned}
$$

$$(1.2)$$

This linear model can be fitted as:

```
m0 <- inla(y ~ s1 + s2, data = SPDEtoy)
```

Although this is a simple example, there are many things going on behind the scenes. Default prior distributions on the intercept, the covariate coefficients, and the precision of the error term have been used. Secondly, the posterior marginals of the model parameters have been approximated using the INLA method and some quantities of interest (such as, for example, the marginal likelihood) have been computed.

A summary of the fitted model can be obtained as:

```
summary(m0)
##
## Call:
##      "inla(formula = y ~ s1 + s2, data = SPDEtoy)"
## Time used:
##      Pre = 2.23, Running = 0.401, Post = 0.174, Total = 2.8
## Fixed effects:
```

```
##                   mean     sd 0.025quant 0.5quant 0.975quant   mode
## (Intercept) 10.13 0.24        9.656      10.13       10.61 10.13
## s1                0.76 0.43     -0.081      0.76        1.61  0.76
## s2               -1.58 0.43     -2.428     -1.58       -0.74 -1.58
##                   kld
## (Intercept)    0
## s1             0
## s2             0
##
## The model has no random effects
##
## Model hyperparameters:
##
##                                               mean     sd 0.025quant
## Precision for the Gaussian observations 0.308 0.031      0.251
##                                             0.5quant 0.975quant
## Precision for the Gaussian observations    0.307      0.372
##                                               mode
## Precision for the Gaussian observations 0.305
##
## Expected number of effective parameters(stdev): 3.00(0.00)
## Number of equivalent replicates : 66.67
##
## Marginal log-Likelihood:  -423.18
```

The output provides a summary of the posterior marginals of the intercept, coefficients of the covariates and the precision of the error term. The posterior marginals of these parameters can be seen in Figure 1.2.

Note that the results provided in the summary of the model and the posterior marginals in Figure 1.2 both suggest that the value of the response decreases as the value of the y-coordinate s2 increases.

1.3.1   Non-linear effects of covariates

As described in Section 1.2, the INLA methodology can handle different types of effects, including random effects. In this case, in order to explore non-linear trends on the coordinates, a non-linear effect can be added on the covariates by using a random walk of order 1. This will replace the fixed effects on the covariates by a smooth term that may represent the effect of the coordinates better.

The vector of random effects $\mathbf{u} = (u_1, \ldots, u_n)$ is defined assuming independent increments:

$$\Delta u_i = u_i - u_{i+1} \sim N(0, \tau_u^{-1}), \ i = 1, \ldots, n - 1$$

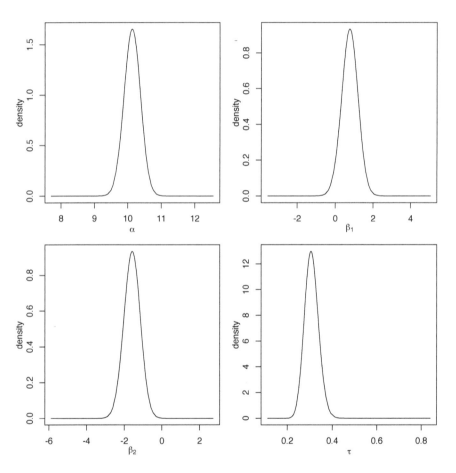

FIGURE 1.2 Posterior marginals of the parameters in the linear model fitted to the SPDEtoy dataset.

Here, τ_u represents the precision of the random walk of order 1. Note that this is a discrete model and that, when it is defined on a continuous covariate (as in this case), the effect associated with observation y_i is $u_{(i)}$. Here, the index (i) is the position of the covariate associated with y_i used to define the model when all the values of the covariate are ordered increasingly.

Hence, the model now is defined using the following mean for the observations:

$$\mu_i = \alpha + u_{1,(i)} + u_{2,(i')}$$

Vectors $\mathbf{u}_1 = (u_{1,1}, \ldots, u_{1,n})$ and $\mathbf{u}_2 = (u_{2,1}, \ldots, u_{2,n})$ represent the random effects associated with the covariates. Indices (i) and (i') are the positions of covariates s_{1i} and s_{2i}, respectively. The precisions of the random effects are τ_1 and τ_2, respectively. By default, these parameters will be assigned a Gamma prior with parameters 1 and 0.00005. Full details about this model are available in the INLA package documentation, which can be viewed with `inla.doc("rw1")`.

Random effects in INLA are defined by using the `f()` function within the formula that defines the model. The model defined before with non-linear terms on the covariates to introduce a smooth term on them can be fitted as:

```
f.rw1 <- y ~ f(s1, model = "rw1", scale.model = TRUE) +
  f(s2, model = "rw1", scale.model = TRUE)
```

In the previous formula, the `f()` function takes two arguments: the value of the covariate, and the type of random effect using the `model` argument. As we have decided to use a random walk of order one, the model is defined as `model = "rw1"`. Furthermore, option `scale.model = TRUE` makes the model to be scaled to have an average variance of 1 (Sørbye and Rue, 2014).

A complete list of implemented random effects can be obtained with `inla.models()$latent`. This will produce a named list with some computational details of the implemented models. A complete list with the names of the implemented models can be obtained with `names(inla.models()$latent)`, or `inla.list.models("latent")`.

Then, the model is fitted and summarized as follows:

```
m1 <- inla(f.rw1, data = SPDEtoy)

summary(m1)
##
## Call:
##    "inla(formula = f.rw1, data = SPDEtoy)"
## Time used:
```

```
##        Pre = 2.68, Running = 0.872, Post = 0.205, Total = 3.76
## Fixed effects:
##                  mean    sd 0.025quant 0.5quant 0.97quant mode kld
## (Intercept)  9.9 0.12        9.6      9.9         10 9.9   0
##
## Random effects:
##    Name      Model
##      s1 RW1 model
##      s2 RW1 model
##
## Model hyperparameters:
##                                              mean       sd
## Precision for the Gaussian observations    0.351     0.04
## Precision for s1                           6.655    18.46
## Precision for s2                          47.089   187.21
##                                          0.025quant 0.5quant
## Precision for the Gaussian observations    0.276     0.35
## Precision for s1                           0.445     2.52
## Precision for s2                           1.378    13.13
##                                          0.97quant   mode
## Precision for the Gaussian observations    0.43    0.350
## Precision for s1                          34.06    0.907
## Precision for s2                         260.36    3.138
##
## Expected number of effective parameters(stdev): 11.63(5.00)
## Number of equivalent replicates : 17.19
##
## Marginal log-Likelihood:   -1169.67
```

Figure 1.3 summarizes the effects fitted in this model. This includes the posterior marginals of the intercept α and the precision of the error term τ, as well as the non-linear effects on the covariates. Note how the effect on the covariates is now slightly non-linear. The posterior marginals of the precisions of both random walks on covariates s1 and s2 (τ_1 and τ_2, respectively) have also been plotted.

The results show a decreasing trend on s2 but an unclear trend on s1. Furthermore, the posterior marginals of the precisions τ_1 and τ_2 show that the effect on covariate s2 has a stronger signal. Although this second model is more complex than the previous one it does not seem to improve model fitting, as the non-linear terms show no clear effect on s1 and a linear decreasing trend on s2. Note that the marginal likelihood cannot be used for the improper models, as the renormalization constant is not included. See inla.doc("rw1") for details.

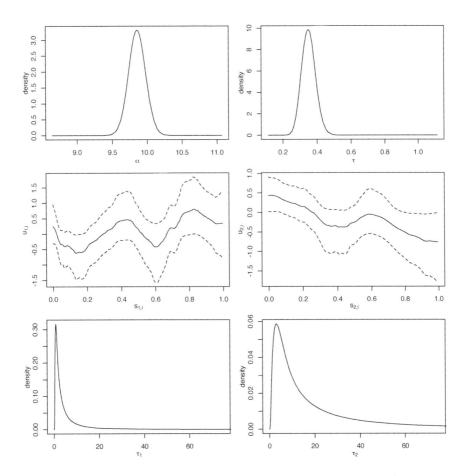

FIGURE 1.3 Posterior marginals of the intercept and precisions of the model with non-linear effects on the covariates fitted to the `SPDEtoy` dataset. The non-linear effects on the covariates are summarized using the posterior mean (solid line) and the limits of 95% credible intervals (dashed line).

1.3.2   inla objects

The object returned by function inla() is of type inla and it is a list that contains all the results from the model fitting with the INLA methodology. The actual results included may depend on the options used in the call to inla() (see Section 1.4 for some more details).

Table 1.2 describes some of the elements available in an inla object. Note that some of them are computed by default but that others (mostly the marginals of some effects and some model assessment criteria) are only computed when the appropriate options are passed on to inla().

TABLE 1.2: Elements in an inla object returned by a call to inla().

Function	Description
summary.fixed	Summary of fixed effects.
marginals.fixed	List of marginals of fixed effects.
summary.random	Summary of random effects.
marginals.random	List of marginals of random effects.
summary.hyperpar	Summary of hyperparameters.
marginals.hyperpar	List of marginals of the hyperparameters.
mlik	Marginal log-likelihood.
summary.linear.predictor	Summary of linear predictors.
marginals.linear.predictor	List of marginals of linear predictors.
summary.fitted.values	Summary of fitted values.
marginals.fitted.values	List of marginals of fitted values.

Summaries in an inla object are typically a data.frame with the estimates of the posterior mean, standard deviation, quantiles (by default, 0.025, 0.5 and 0.975) and the mode. Marginals are represented in named lists by 2-column matrices, where the first column is the value of the parameter and the second column is the density.

For example, in order to display the summary of the fixed values from the linear regression model, we could do the following:

```
m0$summary.fixed
##                mean    sd 0.025quant 0.5quant 0.975quant    mode
## (Intercept) 10.132 0.242     9.6561   10.132      10.61 10.132
## s1           0.762 0.429    -0.0815    0.762       1.61  0.762
## s2          -1.584 0.429    -2.4276   -1.584      -0.74 -1.584
##                kld
## (Intercept) 6.25e-07
```

```
## s1            6.25e-07
## s2            6.25e-07
```

For the fixed effects, `inla()` computes the symmetric Kullback-Leibler divergence using the Gaussian and Laplace approximations, and this is shown under column `kld`.

Similarly, the posterior marginal of the intercept α could be plotted using:

```
plot(m0$marginals.fixed[[1]], type = "l",
  xlab = expression(alpha), ylab = "density")
```

This is the command that has been used to produce the top-left plot in Figure 1.2. Marginals for the other fixed effects and hyperparameters could be plotted in a similar way. Calling `plot()` on an `inla` object will produce a number of default plots as well.

1.3.3 Prediction

Bayesian inference treats missing observations as any other parameter in the model (Little and Rubin, 2002). When the missing observations are in the response variable, INLA automatically computes the predictive distribution of the corresponding linear predictor and fitted values. These missing observations in the response will be assigned the `NA` value in R, so that INLA knows that they are missing values.

In spatial statistics, prediction is often required when dealing with continuous spatial processes as the interest is in estimating the response variable at any point in the study region. In the next example, the predictive distribution of the response variable will be approximated at location $(0.5, 0.5)$.

A new line to the `SPDEtoy` dataset will be added with the `NA` value and the coordinates of the point:

```
SPDEtoy.pred <- rbind(SPDEtoy, c(NA, 0.5, 0.5))
```

Next, the model will be fitted to the newly created `SPDEtoy.pred` dataset, which contains 201 observations. Note that option `compute = TRUE` needs to be set in `control.predictor` in order to compute the posterior marginals of the fitted values. The model with fixed effects on the covariates will be used:

```
m0.pred <- inla(y ~ s1 + s2, data = SPDEtoy.pred,
  control.predictor = list(compute = TRUE))
```

The results now provide the predictive distribution for the missing observation (in the 201st position of the list of marginals of the fitted values):

```
m0.pred$marginals.fitted.values[[201]]
```

This distribution is displayed in Figure 1.4. INLA provides several functions to manipulate the posterior marginals and compute quantities of interest, as explained in Section 1.5. For example, the posterior mean and variance could be easily computed.

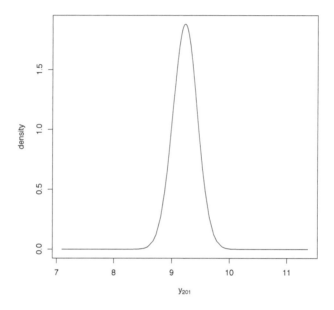

FIGURE 1.4 Predictive distribution of the response at location (0.5, 0.5).

1.4 Additional arguments and control options

As mentioned above, `inla()` can take a number of options that will help to define the model and the way the approximation with the INLA method is computed. Table 1.3 shows a few arguments that are sometimes required when defining a model. Some of them will be used in later sections of this book.

TABLE 1.3: Some arguments taken by `inla()` to define a model and produce a summary of model fitting.

Argument	Description
quantiles	Quantiles to be computed in the summary (default is `c(0.025, 0.5, 0.975)`).
E	Expected values (for some Poisson models, default is `NULL`).
offset	Offset to be added to the linear predictor (default is `NULL`).
weights	Weights on the observations (default is `NULL`)
Ntrials	Number of trials (for some binomial models, default is `NULL`).
verbose	Verbose output (default is `FALSE`).

Table 1.4 shows the main arguments taken by `inla()` to control the estimation process. Note that there are other control arguments not shown here.

TABLE 1.4: Some arguments taken by `inla()` to customize the estimation process.

Argument	Description
control.fixed	Control options for fixed effects.
control.family	Control options for the likelihood.
control.compute	Control options for what is computed (e.g., DIC, WAIC, etc.)
control.predictor	Control options for the linear predictor.
control.inla	Control options for how the posterior is computed.
control.results	Control options for computing the marginals of random effects and linear predictors.
control.mode	Control options to set the modes of the hyperparameters.

All the control arguments must take a named list with different options. These are conveniently described in the associated manual pages and will not be discussed here. For example, to find the list of possible options to be used with `control.fixed`, the user can type `?control.fixed` in the R console.

A typical example of the use of these control options is computing some criteria for model selection and assessment, such as the DIC (Spiegelhalter et al., 2002), WAIC (Watanabe, 2013), CPO (Pettit, 1990), or PIT (Marshall and Spiegelhalter, 2003). These can be computed by setting the appropriate options in `control.compute`:

```
m0.opts <- inla(y ~ s1 + s2, data = SPDEtoy,
  control.compute = list(dic = TRUE, cpo = TRUE, waic = TRUE)
)
```

The output with the new options is similar to the previous one, but now the
DIC and the WAIC are reported:

summary(m0.opts)
```
## 
## Call:
##    c("inla(formula = y ~ s1 + s2, data = SPDEtoy,
##    control.compute = list(dic = TRUE, ", " cpo = TRUE,
##    waic = TRUE))")
## Time used:
##    Pre = 2.51, Running = 1.14, Post = 0.251, Total = 3.91
## Fixed effects:
##               mean   sd 0.025quant 0.5quant 0.975quant   mode
## (Intercept) 10.13 0.24      9.656    10.13      10.61  10.13
## s1           0.76 0.43     -0.081     0.76       1.61   0.76
## s2          -1.58 0.43     -2.428    -1.58      -0.74  -1.58
##               kld
## (Intercept)    0
## s1             0
## s2             0
## 
## The model has no random effects
## 
## Model hyperparameters:
##                                             mean    sd 0.025quant
## Precision for the Gaussian observations    0.308 0.031      0.251
##                                          0.5quant 0.975quant
## Precision for the Gaussian observations     0.307      0.372
##                                             mode
## Precision for the Gaussian observations    0.305
## 
## Expected number of effective parameters(stdev): 3.00(0.00)
## Number of equivalent replicates : 66.67
## 
## Deviance Information Criterion (DIC) ...............: 810.09
## Deviance Information Criterion (DIC, saturated) ....: 207.16
## Effective number of parameters .....................: 4.08
## 
## Watanabe-Akaike information criterion (WAIC) ...: 809.78
```

```
## Effective number of parameters ................: 3.69
##
## Marginal log-Likelihood:   -423.18
## CPO and PIT are computed
##
## Posterior marginals for the linear predictor and
##  the fitted values are computed
```

The CPO and PIT are not reported, but a message in the output states that they have been computed. They can be accessed as:

```
m0.opts$cpo$cpo
m0.opts$cpo$pit
```

Figure 1.5 shows histograms of the computed CPO and PIT values for the model with fixed effects.

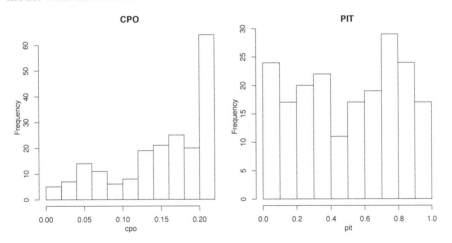

FIGURE 1.5 Histograms of CPO and PIT values for the model with fixed effects.

1.4.1   Estimation method

Other important options are related to the method used by INLA to approximate the posterior distribution of the hyperparameters. These are controlled by `control.inla` and some of them are summarized in Table 1.5.

The estimation method can be changed to achieve higher accuracy in the approximations or reduce the computation time by resorting to less accurate approximations. For example, the Gaussian approximation in INLA is less computationally expensive than the Laplace approximation.

Similarly, the integration strategy can be set in different ways. INLA uses a central composite design (CCD, Box and Draper, 2007) to estimate the posterior distribution of the hyperparameters. This can be changed to an empirical Bayes (EB, Carlin and Louis, 2008) strategy so that no integration of the hyperparameters is done. Other integration strategies are available, as listed in Table 1.5 and documented in the manual page of `control.inla`.

For example, we could compute the model with non-linear random effects using two different integration strategies and then compare the computation times. In particular, CCD and EB integration strategies will be compared:

```
m1.ccd <- inla(f.rw1, data = SPDEtoy,
   control.compute = list(dic = TRUE, cpo = TRUE, waic = TRUE),
   control.inla = list(int.strategy = "ccd"))
m1.eb <- inla(f.rw1, data = SPDEtoy,
   control.compute = list(dic = TRUE, cpo = TRUE, waic = TRUE),
   control.inla = list(int.strategy = "eb"))
```

TABLE 1.5: Some options that can be passed though `control.inla` to control the estimation process with INLA. Check the manual page (with `?control.inla`) for more details.

Option	Description
`strategy`	Strategy used for the approximations: `simplified.laplace` (default), `adaptive`, `gaussian` or `laplace`.
`int.strategy`	Integration strategy: `auto` (default), `ccd`, `grid` or `eb` (check manual page for other options).

As seen below, the empirical Bayes integration strategy `eb` fits the model faster, but will likely provide less accurate approximations to the posterior marginals. In this example, the time difference is minimal, but for more complex models we can get massive speed-up from using `eb`.

```
# CCD strategy
m1.ccd$cpu.used
##      Pre Running    Post    Total
## 2.6528  0.9115  0.1861  3.7505

# EB strategy
```

```
m1.eb$cpu.used
##      Pre Running     Post    Total
##    2.0328  0.7252  0.1626   2.9205
```

1.4.2   Setting the priors

INLA has a set of predefined priors for all the effects in the model that will be used by default. However, prior specification is a key step in Bayesian analysis and careful attention should be paid to the specification of the priors in the model.

For the fixed effects, the default prior is a Gaussian distribution with zero mean, and the precision is zero for the intercept and 0.001 for the coefficients of the covariates. These values for the Gaussian prior can be changed with argument control.fixed using the variables listed in Figure 1.6. More information is available in the manual page (which can be accessed with ?control.fixed).

TABLE 1.6: Options to set the prior of the fixed effects in argument control.fixed.

Option	Description
mean.intercept	Prior mean for the intercept (default is 0).
prec.intercept	Prior precision for the intercept (default is 0).
mean	Prior mean for the coefficients of the covariates (default is 0). It can be a named list.
prec	Prior precision for the coefficients of the covariates (default is 0.001). It can be a named list.

The parameters in the likelihood can also be assigned a prior by means of variable hyper in argument control.likelihood. Similarly, the hyperparameters of the random effects defined through the f() function can be assigned a prior using argument hyper. In both cases, hyper will be a named list (using the names of the hyperparameters) and each value will be a list with the options listed in Table 1.7.

TABLE 1.7: Options to set the prior of the hyperparameters in the likelihood and random effects.

Option	Description
initial	Initial value of the hyperparameter.
prior	Prior distribution to be used.
param	Vector with the values of the parameters of the prior distribution.

Option	Description
fixed	Boolean variable to set the parameter to a fixed value (default FALSE).

In order to show how the different priors are defined, we will fit the model with non-linear effects again. Note that all priors are set in the internal scale of the parameters in INLA. This is reported in the package documentation and, for example, the precision is represented internally in the log-scale, so that the prior is set on the log-precision.

First of all, fixed effects will be set to have a Gaussian prior with zero mean and precision 1. These are the options that will be passed using argument control.fixed:

```
# Prior on the fixed effects
prior.fixed <- list(mean.intercept = 0, prec.intercept = 1,
  mean = 0, prec = 1)
```

Similarly, the log-precision in the Gaussian likelihood of the model will be assigned a Gaussian prior with zero mean and precision 1, and this will be passed using argument control.family:

```
# Prior on the likelihood precision (log-scale)
prior.prec <- list(initial = 0, prior = "normal", param = c(0, 1),
  fixed = FALSE)
```

Note how the initial value is set to zero as it is in the internal scale; i.e., the initial value of the precision is $\exp(0) = 1$.

Finally, the precision of the random walks will be fixed to 1 (i.e., 0 in the internal scale):

```
# Prior on the precision of the RW1
prior.rw1 <- list(initial = 0, fixed = TRUE)
```

This means that the hyperparameter is not estimated but fixed to the value provided and, hence, it will not be reported in the output.

The following R code shows the call to fit the model using the different sets of priors. Note how the priors on the likelihood hyperparameters and the random effects need to be embedded in a named list using the name of the hyperparameter:

```
f.hyper <- y ~ 1 +
  f(s1, model = "rw1", hyper = list(prec = prior.rw1),
    scale.model = TRUE) +
  f(s2, model = "rw1", hyper = list(prec = prior.rw1),
    scale.model = TRUE)

m1.hyper <- inla(f.hyper, data = SPDEtoy,
  control.fixed = prior.fixed,
  control.family = list(hyper = list(prec = prior.prec)))
```

A summary of the model can be obtained as follows:

```
summary(m1.hyper)
##
## Call:
##    c("inla(formula = f.hyper, data = SPDEtoy,
##    control.family = list(hyper = list(prec = prior.prec)),
##    ", " control.fixed = prior.fixed)")
## Time used:
##    Pre = 2.09, Running = 0.391, Post = 0.129, Total = 2.61
## Fixed effects:
##              mean    sd 0.025quant 0.5quant 0.975quant   mode
## (Intercept) 9.726 0.117      9.494    9.726      9.953 9.728
##              kld
## (Intercept)    0
##
## Random effects:
##    Name     Model
##      s1 RW1 model
##      s2 RW1 model
##
## Model hyperparameters:
##                                          mean    sd 0.025quant
## Precision for the Gaussian observations 0.37 0.039      0.298
##                                        0.5quant 0.975quant
## Precision for the Gaussian observations   0.369      0.452
##                                          mode
## Precision for the Gaussian observations 0.365
##
## Expected number of effective parameters(stdev): 20.46(1.08)
## Number of equivalent replicates : 9.77
##
## Marginal log-Likelihood:  -1193.32
```

As mentioned above, the precisions of the non-linear terms on the covariates are not reported because they have been fixed to 1.

A complete list of the priors available in the INLA package (and their options) can be obtained with `inla.models()$prior`. This is a named list, and the names of the priors can be easily checked with `names(inla.models()$prior)`. These are also described in the R-INLA package documentation. An alternative, is to use `inla.list.models("prior")`. Additionally, there is an option for user defined priors through tables or R expressions.

1.5   Manipulating the posterior marginals

The INLA package comes with a number of functions to manipulate the posterior marginals returned by the `inla()` function. These are summarized in Table 1.8 and specific information can be found in their respective manual pages.

TABLE 1.8: Functions to manipulate the posterior marginals.

Function	Description
inla.emarginal()	Compute the expectation of a function.
inla.dmarginal()	Compute the density.
inla.pmarginal()	Compute a probability.
inla.qmarginal()	Compute a quantile.
inla.rmarginal()	Sample from the marginal.
inla.hpdmarginal()	Compute a high probability density (HPD) interval.
inla.smarginal()	Interpolate the posterior marginal.
inla.mmarginal()	Compute the mode.
inla.tmarginal()	Transform the marginal.
inla.zmarginal()	Compute summary statistics.

A typical example of the use of these functions is to compute the posterior marginal of the variance of the error term (i.e., $1/\tau$). This would involve a transformation of the posterior marginal on τ, and then computing some summary statistics.

The transformation can be done using function `inla.tmarginal()`, which will take the function for the transformation and the marginal. We strongly recommend transforming the `internal.marginals.hyperpar` instead of the `marginals.hyperpar`. The `internal.marginals.hyperpar` reports the marginals of the hyperparameters in the internal scale, like $\log(\tau)$ in this example:

```
# Compute posterior marginal of variance
post.var <- inla.tmarginal(function(x) exp(-x),
  m0$internal.marginals.hyperpar[[1]])
```

Summary statistics can be computed with function `inla.zmarginal()`:

```
# Compute summary statistics
inla.zmarginal(post.var)
## Mean            3.27668
## Stdev           0.329579
## Quantile  0.025 2.69202
## Quantile  0.25  3.0438
## Quantile  0.5   3.25438
## Quantile  0.75  3.4848
## Quantile  0.975 3.98493
```

Similarly, a 95% high probability interval (HPD) can be computed as:

```
inla.hpdmarginal(0.95, post.var)
##               low   high
## level:0.95 2.655 3.936
```

1.6 Advanced features

In this section we will describe some advanced features available in the `INLA` package for model fitting. In particular, the use of the models with more than one likelihood, how to define models that share effects between different parts of the model, how to create linear combinations on the latent effects and the use of Penalized Complexity priors. More details about advanced features can be found in the R-INLA website at `http://www.r-inla.org/models/tools`.

1.6.1 Several likelihoods

`INLA` can handle models with more than one likelihood. This is a common modeling approach to build joint models. For example, in survival analysis survival times may be related to some outcome of interest that may be modeled jointly using a longitudinal model (Ibrahim et al., 2001). By using a model with more than one likelihood it is possible to build a joint model with different types of outputs and the hyperparameters in the likelihoods will be fitted separately.

In order to provide a simple example, we will consider a similar toy dataset to the SPDEtoy one with added noise. This would mimic a situation in which two sets of measurements are available but with different noise effects. This new dataset can be created by adding random Gaussian noise (with standard deviation equal to 2) to the original observations in the SPDEtoy dataset:

```
library(INLA)
data(SPDEtoy)
SPDEtoy$y2 <- SPDEtoy$y + rnorm(nrow(SPDEtoy), sd = 2)
```

Hence, both sets of observations can be modeled on the same covariates but the precisions of the error terms are different, and these should be modeled separately. A simple way to do this is to fit a model with two Gaussian likelihoods with different precisions to be estimated by each one of the sets of observations.

The model that will be fitted is the following:

$$
\begin{aligned}
y_i &\sim N(\mu_i, \tau_1^{-1}), \ i = 1, \ldots, 200 \\
y_i &\sim N(\mu_i, \tau_2^{-1}), \ i = 201, \ldots, 400 \\
\mu_i &= \alpha + \beta_1 s_{1i} + \beta_2 s_{2i}, \ i = 1, \ldots, 400 \\
\alpha &\sim \text{Uniform} \\
\beta_j &\sim N(0, 0.001^{-1}), \ j = 1, 2 \\
\tau_j &\sim Ga(1, 0.00005), \ j = 1, 2
\end{aligned}
$$

$$(1.3)$$

In order to fit a model with more than one likelihood the response variable must be a matrix with as many columns as the number of likelihoods. The number of rows is the total number of observations. Given a column, the rows with data not associated with that likelihood are filled with NA values. In our example, this can be done as follows:

```
# Number of locations
n <- nrow(SPDEtoy)

# Response matrix
Y <- matrix(NA, ncol = 2, nrow = n * 2)

# Add `y` in first column, rows 1 to 200
Y[1:n, 1] <- SPDEtoy$y
# Add `y2` in second column, rows 201 to 400
Y[n + 1:n, 2] <- SPDEtoy$y2
```

Note how the values of y are added in the first column in rows 1 to n, and the values of y2 are added to the second column in rows from n + 1 to 2 * n, where n is the number of locations in the SPDEtoy dataset.

Using more than one likelihood may affect other elements in the model definition, such as if the model incorporates in the likelihood some of the values described in Table 1.3.

The covariates will be the same in both likelihoods, so they do not need to be modified in any way. The family argument must be a vector with the names of the likelihoods used. In this case, it will be family = c("gaussian", "gaussian") but different likelihoods can be used.

Then, the model can be fitted as follows:

```
m0.2lik <- inla(Y ~ s1 + s2, family = c("gaussian", "gaussian"),
  data = data.frame(Y = Y,
    s1 = rep(SPDEtoy$s1, 2),
    s2 = rep(SPDEtoy$s2, 2))
)
```

The summary of this model now shows estimates of two precisions:

```
summary(m0.2lik)
##
## Call:
##    c("inla(formula = Y ~ s1 + s2, family = c(\"gaussian\",
##    \"gaussian\"), ", " data = data.frame(Y = Y, s1 =
##    rep(SPDEtoy$s1, 2), s2 = rep(SPDEtoy$s2, ", " 2)))")
## Time used:
##    Pre = 1.93, Running = 0.501, Post = 0.151, Total = 2.58
## Fixed effects:
##                mean    sd 0.025quant 0.5quant 0.975quant   mode
## (Intercept) 10.22 0.199      9.827    10.22     10.608  10.22
## s1           0.64 0.352     -0.053     0.64      1.331   0.64
## s2          -1.57 0.352     -2.264    -1.57     -0.881  -1.57
##                kld
## (Intercept)    0
## s1             0
## s2             0
##
## The model has no random effects
##
## Model hyperparameters:
##                                              mean      sd
```

```
## Precision for the Gaussian observations     0.310 0.031
## Precision for the Gaussian observations[2] 0.147 0.015
##                                               0.025quant 0.5quant
## Precision for the Gaussian observations           0.253    0.308
## Precision for the Gaussian observations[2]         0.120    0.146
##                                               0.975quant   mode
## Precision for the Gaussian observations           0.374 0.306
## Precision for the Gaussian observations[2]        0.177 0.145
##
## Expected number of effective parameters(stdev): 3.00(0.00)
## Number of equivalent replicates : 133.32
##
## Marginal log-Likelihood:   -913.26
```

Note how the precision of the second likelihood has a posterior mean smaller to the precision of the first likelihood. This is because the variance of the second set of observations is higher as they were generated by adding some noise to the original data. Also, the estimates of the fixed effects are very similar to the ones estimated in the previous models.

1.6.2   *Copy* model

Sometimes it is necessary to share an effect that is estimated from two or more parts of the dataset, so that all of them provide information about the effect when fitting the model. This is known as a *copy* effect, as the new effect will be a copy of the original effect plus some tiny noise.

To illustrate the use of the *copy* effect in INLA we will fit a model where the fixed effect of the y-coordinate is copied. In particular, this will be the fitted model:

$$
\begin{aligned}
y_i &\sim N(\mu_i, \tau^{-1}), \; i = 1, \ldots, 400 \\
\mu_i &= \alpha + \beta_1 s_{1i} + \beta_2 s_{2i}, \; i = 1, \ldots, 200 \\
\mu_i &= \alpha + \beta_1 s_{1i} + \beta \cdot \beta_2^* s_{2i}, \; i = 201, \ldots, 400 \\
\alpha &\sim \text{Uniform} \\
\beta_j &\sim N(0, 0.001^{-1}), \; j = 1, 2 \\
\beta_2^* &\sim N(\beta_2, \tau_{\beta_2}^{-1} = 1/\exp(14)) \\
\tau_j &\sim Ga(1, 0.00005), \; j = 1, 2
\end{aligned}
$$

Here, β_2^* is the copied effect (from β_2). Note how the copied effect has a scaling factor, β, but this is fixed to 1 by default. Furthermore, the precision of β_2^* is set to a very large value, ensuring that the copied effect is very close to β_2.

This precision can be set using argument `precision` in the call of the `f()` function (see below).

Before implementing the model in `INLA`, a new `data.frame` with the original `SPDEtoy` data and the simulated one will be put together. This will involve creating two new vectors of covariates by repeating the original covariates. Furthermore, two indices will be created to identify to which group of observations a value belongs:

```
y.vec <- c(SPDEtoy$y, SPDEtoy$y2)
r <- rep(1:2, each = nrow(SPDEtoy))
s1.vec <- rep(SPDEtoy$s1, 2)
s2.vec <- rep(SPDEtoy$s2, 2)
i1 <- c(rep(1, n), rep(NA, n))
i2 <- c(rep(NA, n), rep(1, n))

d <- data.frame(y.vec, s1.vec, s2.vec, i1, i2)
```

The copied effect will be defined using independent and identically distributed random effects with a single group that will be multiplied by the values of the covariates. This is an alternative way to define a linear effect in `INLA` in a way that allows the effect to be copied. This model is implemented as the `iid` model in `INLA`. See `inla.doc("iid")` for details.

For this reason, indices `i1` and `i2` have either 1 (i.e., the linear predictor includes the random effect) or `NA` (there is no random effect in the linear predictor). This ensures that the linear predictor includes the original random effect in the first 200 observations (using index `i1`), and the copied effect in the last 200 observations (using index `i2`).

Given that the model is now implemented using a single random effect with the covariate values as weights, the precision of the random effect needs to be fixed to `0.001` in order to obtain similar results as in previous models. This can be done by using an initial value for the precision of the random effects and fixing it in the prior definition. The values to be passed to the prior definition (in the call to the `f()` function below) are:

```
tau.prior = list(prec = list(initial = 0.001, fixed = TRUE))
```

Then, the formula to fit the model is defined using first an `iid` model with an index `i1` which is 1 for the first `n` values and `NA` for the rest. The copied effect uses an index for the second set of observations, and the values of the covariates as weights:

```
f.copy <- y.vec ~ s1.vec +
  f(i1, s2.vec, model = "iid", hyper = tau.prior) +
  f(i2, s2.vec, copy = "i1")
```

Finally, the model is fitted and summarized as:

```
m0.copy <- inla(f.copy, data = d)
```

```
summary(m0.copy)
##
## Call:
##     "inla(formula = f.copy, data = d)"
## Time used:
##     Pre = 2.59, Running = 0.356, Post = 0.136, Total = 3.09
## Fixed effects:
##                  mean     sd 0.025quant 0.5quant 0.975quant    mode
## (Intercept) 10.197 0.208      9.788   10.198      10.61 10.198
## s1.vec       0.579 0.378     -0.164    0.578       1.32  0.578
##                  kld
## (Intercept)    0
## s1.vec         0
##
## Random effects:
##    Name       Model
##       i1 IID model
##       i2 Copy
##
## Model hyperparameters:
##                                                 mean     sd 0.025quant
## Precision for the Gaussian observations 0.198 0.014      0.171
##                                               0.5quant 0.975quant
## Precision for the Gaussian observations    0.198      0.226
##                                                 mode
## Precision for the Gaussian observations 0.197
##
## Expected number of effective parameters(stdev): 2.88(0.008)
## Number of equivalent replicates : 139.12
##
## Marginal log-Likelihood:   -912.05
```

Note that with this particular parameterization of the model, the coefficient of
s2 is in fact a random effect and that it is available in the summary of the
random effects:

```
m0.copy$summary.random
## $i1
##    ID   mean     sd 0.025quant 0.5quant 0.975quant   mode      kld
## 1   1 -1.37 0.354      -2.06    -1.37     -0.675 -1.37 1.94e-07
##
## $i2
##    ID   mean     sd 0.025quant 0.5quant 0.975quant   mode      kld
## 1   1 -1.37 0.354      -2.06    -1.37     -0.675 -1.37 1.94e-07
```

Alternatively, the copied effect can also be multiplied by a scaling factor β, which is estimated. This can be achieved by adding `fixed = FALSE` to the definition of the copied effect:

```
f.copy2 <- y.vec ~ s1.vec + f(i1, s2.vec, model = "iid") +
  f(i2, s2.vec, copy = "i1", fixed = FALSE)
```

In this case, the estimate of this coefficient should be close to one as the copied effect is exactly the same for the second group of observations:

```
m0.copy2 <- inla(f.copy2, data = d)
summary(m0.copy2)
##
## Call:
##    "inla(formula = f.copy2, data = d)"
## Time used:
##     Pre = 2.51, Running = 0.806, Post = 0.128, Total = 3.44
## Fixed effects:
##               mean    sd 0.025quant 0.5quant 0.975quant   mode
## (Intercept) 9.726 0.172      9.388    9.726     10.064 9.726
## s1.vec      0.623 0.385     -0.134    0.623      1.379 0.623
##               kld
## (Intercept)   0
## s1.vec        0
##
## Random effects:
##    Name      Model
##     i1 IID model
##     i2 Copy
##
## Model hyperparameters:
##                                             mean        sd
## Precision for the Gaussian observations     0.19 1.30e-02
## Precision for i1                        18856.30 1.85e+04
```

```
## Beta for i2                                   1.00 3.16e-01
##                                        0.025quant 0.5quant
## Precision for the Gaussian observations    0.165     0.19
## Precision for i1                        1235.944 13407.56
## Beta for i2                                  0.379     1.00
##                                        0.975quant     mode
## Precision for the Gaussian observations   2.18e-01    0.189
## Precision for i1                         6.78e+04 3334.776
## Beta for i2                               1.62e+00    0.999
##
## Expected number of effective parameters(stdev): 2.00(0.001)
## Number of equivalent replicates : 199.89
##
## Marginal log-Likelihood:   -918.50
```

In the previous output, `Beta for i2` refers to the scaling coefficient β that multiplies the copied effect. It can be seen how its posterior mean is very close to one, as expected.

1.6.3 *Replicate* model

The *copy* effect described in the previous section is useful to create an effect which is very close to the copied one. The *replicate* effect in INLA is similar to the *copy* effect but, in this case, the values of the hyperparameters will be shared by the different replicated effects. This means that, for example, if an `iid` effect is replicated, the replicated random effects will be independent with the same precision.

To show the use of the *replicate* feature in INLA the `linear` latent model will be used. This is essentially an alternative implementation of a linear fixed effect that can be defined through the `f()` function. The only hyperparameter in this effect is the coefficient of the linear effect, which is assigned (by default) a Gaussian prior with zero mean and precision 0.001. These values can be changed in the call to `f()` by setting arguments `mean.linear` and `prec.linear`, respectively. If these are not set, the default values are taken from the settings of `control.fixed`. See `inla.doc("linear")` for full details about this latent effect.

By replicating a `linear` effect, the same coefficient is used in all the replicated effects, which is essentially the same as using the same coefficient. The index to create the replicated effects is created by defining a vector with two values as follows:

```
d$r <- rep(1:2, each = nrow(SPDEtoy))
```

Index r defines how the observations are grouped into the replicated effects. In this case, the first 200 observations are in the first group and the last 200 in the second group.

Then, the linear effects can be replicated in the formula that defines the model as:

```
f.rep <- y.vec ~ f(s1.vec, model = "linear", replicate = r) +
  f(s2.vec, model = "linear", replicate = r)
```

Finally, the model with the replicated effects can be easily fitted:

```
m0.rep <- inla(f.rep, data = d)
```

```
summary(m0.rep)
##
## Call:
##    "inla(formula = f.rep, data = d)"
## Time used:
##    Pre = 1.57, Running = 0.318, Post = 0.11, Total = 2
## Fixed effects:
##                 mean     sd 0.025quant 0.5quant 0.975quant     mode
## (Intercept) 10.265 0.213      9.846   10.265     10.683   10.265
## s1.vec       0.572 0.378     -0.170    0.572      1.314    0.572
## s2.vec      -1.566 0.378     -2.308   -1.566     -0.824   -1.566
##              kld
## (Intercept)   0
## s1.vec        0
## s2.vec        0
##
## The model has no random effects
##
## Model hyperparameters:
##                                               mean     sd 0.025quant
## Precision for the Gaussian observations 0.198 0.014      0.172
##                                          0.5quant 0.975quant
## Precision for the Gaussian observations    0.198      0.226
##                                              mode
## Precision for the Gaussian observations 0.197
##
## Expected number of effective parameters(stdev): 3.00(0.00)
```

```
## Number of equivalent replicates : 133.32
##
## Marginal log-Likelihood:  -914.36
```

Note how the summaries of the effects and hyperparameters are very similar to those obtained in previous examples.

1.6.4   Linear combinations of the latent effects

INLA can also handle models in which the linear predictors defined in the model formula are multiplied by a matrix A (called the *observation matrix* in the package documentation). This is, the vector of linear predictors on the fixed and random effects η defined by the model formula becomes

$$\eta^{*\top} = A\eta^{\top}$$

Here, η^* is the actual linear predictor used when fitting the model and it is a linear combination of the effects in η. Note that η is the linear predictor as defined by the model formula passed to inla().

In order to provide a simple example, we define A as a diagonal matrix with values of 10 in the diagonal. This is similar to multiplying all the effects in the linear predictor by 10, which will make the estimates of the intercept and the covariate coefficients shrink by a factor of 10. This is done as follows:

```
# Define A matrix
A = Diagonal(n + n, 10)

# Fit model
m0.A <- inla(f.rep, data = d, control.predictor = list(A = A))
```

The summary of the fitted model can be seen below. Note how the estimates of the fixed effects are shrunk by a factor of 10 but that the estimate of the precision does not change. This is because the A matrix only affects the linear predictor and not the error term.

```
summary(m0.A)
##
## Call:
##    "inla(formula = f.rep, data = d, control.predictor =
##    list(A = A))"
## Time used:
##    Pre = 1.63, Running = 0.374, Post = 0.111, Total = 2.12
```

```
## Fixed effects:
##              mean      sd 0.025quant 0.5quant 0.975quant    mode
## (Intercept)  1.026 0.021      0.985    1.026      1.068   1.026
## s1.vec       0.057 0.038     -0.017    0.057      0.131   0.057
## s2.vec      -0.157 0.038     -0.231   -0.157     -0.082  -0.157
##              kld
## (Intercept)   0
## s1.vec        0
## s2.vec        0
##
## The model has no random effects
##
## Model hyperparameters:
##                                        mean      sd 0.025quant
## Precision for the Gaussian observations 0.198 0.014      0.172
##                                        0.5quant 0.975quant
## Precision for the Gaussian observations   0.198      0.226
##                                        mode
## Precision for the Gaussian observations 0.197
##
## Expected number of effective parameters(stdev): 3.05(0.003)
## Number of equivalent replicates : 131.22
##
## Marginal log-Likelihood:  -921.27
```

Although this is a very simple example, the A matrix can be useful when the linear predictors in η need to be linearly combined to define more complex linear predictors, such as to allow the effects of the covariates of one observation to influence other observations.

The use of the A matrix plays an important role in the spatial models described through this book. In particular, the A matrix allows us to include linear combinations of some random effects in the model that are needed to define some spatial models. Given the complexity of manually constructing an A matrix from the different effects, we will use helper functions like `inla.stack()` to create this matrix.

1.6.5 Penalized Complexity priors

Simpson et al. (2017) describe a new approach for constructing prior distributions called Penalized Complexity priors, or PC priors for short. Under this new framework, a PC prior for the standard deviation σ of a latent effect is set by defining parameters (u, α) so that

$$\text{Prob}(\sigma > u) = \alpha,\ u > 0,\ 0 < \alpha < 1.$$

Hence, PC priors provide a different way to propose priors on the model hyperparameters.

Section 1.3.1 shows an example of fitting a model with two random walks of order one on s1 and s2 with default Gamma priors on the precision parameters. To set a PC prior on the standard deviation of the random walk, we consider the complexity of the latent effect; e.g., for this effect we believe that the probability of the standard deviation being higher than 1 is quite small. Hence, we set $u = 1$ and $\alpha = 0.01$.

The PC-prior on the standard deviation of the random walk latent effect is then defined as:

```
pcprior <- list(prec = list(prior = "pc.prec",
  param = c(1, 0.01)))
```

This definition will be passed to the f() function when the latent effect is defined in the model formula:

```
f.rw1.pc <- y ~
  f(s1, model = "rw1", scale.model = TRUE, hyper = pcprior) +
  f(s2, model = "rw1", scale.model = TRUE, hyper = pcprior)
```

Next, the model is fitted and the resulting estimates shown:

```
m1.pc <- inla(f.rw1.pc, data = SPDEtoy)

summary(m1.pc)
##
## Call:
##    "inla(formula = f.rw1.pc, data = SPDEtoy)"
## Time used:
##     Pre = 2.22, Running = 0.785, Post = 0.131, Total = 3.13
## Fixed effects:
##                  mean    sd 0.025quant 0.5quant 0.975quant   mode
## (Intercept) 9.858 0.12      9.623    9.858      10.09 9.858
##                  kld
## (Intercept)    0
##
## Random effects:
##    Name       Model
##      s1 RW1 model
```

```
##     s2 RW1 model
##
## Model hyperparameters:
##                                          mean      sd 0.025quant
## Precision for the Gaussian observations 0.354 0.039      0.281
## Precision for s1                        3.771 4.817      0.599
## Precision for s2                        8.055 9.407      1.138
##                                        0.5quant 0.975quant
## Precision for the Gaussian observations   0.353      0.436
## Precision for s1                          2.332     15.689
## Precision for s2                          5.236     32.212
##                                          mode
## Precision for the Gaussian observations 0.351
## Precision for s1                        1.225
## Precision for s2                        2.658
##
## Expected number of effective parameters(stdev): 12.95(3.76)
## Number of equivalent replicates : 15.45
##
## Marginal log-Likelihood:   -1157.66
```

The estimates of the intercept and the precision of the Gaussian likelihood are very similar to the model fitted in Section 1.3.1. The estimates of the precisions of the random walks associated with s1 and s2 seem to be different. In particular, the precision of the random walk on s2 is certainly smaller than when default Gamma priors are used.

In order to inspect the effect of the PC prior we have computed the posterior marginals of the standard deviations of both random walks:

```
post.sigma.s1 <- inla.tmarginal(function (x) sqrt(1 / exp(x)),
  m1.pc$internal.marginals.hyperpar[[2]])

post.sigma.s2 <- inla.tmarginal(function (x) sqrt(1 / exp(x)),
  m1.pc$internal.marginals.hyperpar[[3]])
```

Figure 1.6 shows the posterior marginals of the standard deviations. The PC prior makes the posterior have most of its density below one. In this particular case there are some differences between using a default Gamma prior and a PC prior with parameters $(1, 0.01)$.

More information about the PC prior used in this example can be found in the manual page that can be accessed by typing inla.doc("pc.prec") in R.

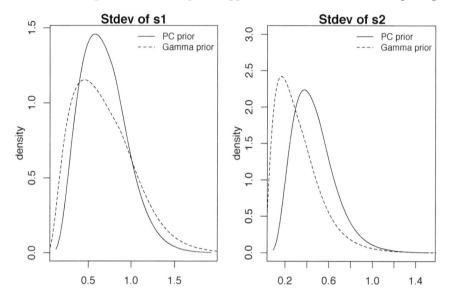

FIGURE 1.6 Posterior marginals of the standard deviations of two random walks using a PC prior and the default Gamma prior.

Other PC priors will be introduced later in the book to propose priors for other types of spatial latent effects.

Section 2.8.2 shows the use of PC priors in a spatial model. In this case, the relevant parameters are the standard deviation of the spatial process and the range, which measures a distance at which spatial autocorrelation is small. PC priors in this case are defined using the same principle as above for both parameters. For the standard deviation σ we need to define (σ_0, α) so that

$$P(\sigma > \sigma_0) = \alpha,$$

while for the range r we need to define (r_0, α) so that

$$P(r < r_0) = \alpha.$$

2

Introduction to spatial modeling

2.1 Introduction

In this section, a brief summary of continuous spatial processes will be provided, and the Stochastic Partial Differential Equations - SPDE approach - as proposed in Lindgren et al. (2011) will be summarized intuitively. Provided that a Gaussian spatial process with Matérn covariance is a solution to SPDE's of the form presented in Lindgren et al. (2011), we will give an intuitive presentation of the main results of this link. The Matérn covariance function is probably the most used one in geostatistics. Therefore, the approach in Lindgren et al. (2011) is useful because the Finite Element Method (FEM) was used to build a GMRF (Rue and Held, 2005) representation and is available for practitioners through the `INLA` package. A recent review of spatial models in INLA can be found in Bakka et al. (2018).

2.1.1 Spatial variation

A point-referenced dataset is made up of any data measured at known locations. These locations may be in any coordinate reference system, most often longitude and latitude. Point-referenced data are common in many areas of science. This type of data appears in mining, climate modeling, ecology, agriculture and other fields. If the influence of location needs to be incorporated into a model, then a model for geo-referenced data is required.

As shown in Chapter 1, a regression model can be built using the coordinates of the data as covariates to describe the spatial variation of the variable of interest. In many cases, a non-linear function based on the coordinates will be necessary to adequately describe the effect of the location. For example, basis functions on the coordinates can be used as covariates in order to build a smooth function. This specification explicitly models the spatial trend on the mean.

Alternatively, it may be more natural to explicitly model the variation of the outcome considering that it may be similar at nearby locations. The first law of geography asserts: "Everything is related to everything else, but near things are more related than distant things" (Tobler, 1970). This means that the proposed model should have the property that an observation is more correlated with an observation close in space than with another observation that has been collected further away.

In spatial statistics it is common to formulate mixed-effects regression models in which the linear predictor is made of a trend plus a spatial variation (Haining, 2003). The trend usually is composed of fixed effects or some smooth terms on covariates, while the spatial variation is usually modeled using correlated random effects. Spatial random effects often model (residual) small scale variation and this is the reason why these models can be regarded as models with correlated errors.

Models that account for spatial dependence may be defined according to whether locations are areas (cities or countries, for example) or points. For a review of models for different types of spatial data see, for example, Haining (2003) and Cressie and Wikle (2011).

The case of areal data (also known as lattice data) is often tackled by the use of generalized linear mixed-effects models given that variables are measured at a discrete number of locations. Given a vector of observations $\mathbf{y} = (y_1, \ldots, y_n)$ from n areas, the model can be represented as

$$y_i | \mu_i, \theta \sim f(\mu_i; \theta)$$

$$g(\mu_i) = X_i \beta + u_i; \ i = 1, \ldots, n$$

Here, $f(\mu_i; \theta)$ is a distribution with mean μ_i and hyperparameters θ. The function $g(\cdot)$ links the mean of each observation to a linear predictor on a vector of covariates X_i with coefficients β. The terms u_i represent random effects, distributed usually using a Gaussian multivariate distribution with zero mean and precision matrix τQ.

While τ is a precision hyperparameter, the structure of the precision matrix is given by the matrix Q and is defined in a way that captures the spatial correlation in the data. For lattice data, Q is often based on the *adjacency* matrix that represents the neighborhood structure of the areas where data has been collected. This adjacency is illustrated in Figure 2.1, where the adjacency structure of the 100 counties in North Carolina is represented by a graph. This graph can be represented using a 100×100 matrix W with 1 at entry (i, j) if counties i and j are neighbors and 0 otherwise.

A simple way to define spatially correlated random effects using the adjacency matrix W is to take $Q = I - \rho W$, where I is the identity matrix and ρ is a spatial autocorrelation parameter. This is known as a CAR specification (Schabenberger and Gotway, 2004). Nevertheless, there are other ways to exploit W to define Q to model spatial autocorrelation. See, for example, Haining (2003), Cressie and Wikle (2011) or Banerjee et al. (2014).

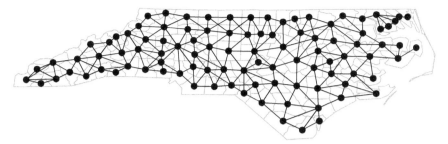

FIGURE 2.1 Counties in North Carolina and their neighborhood structure.

Spatial data that is observed at specific locations is usually divided in two particular cases, depending on whether locations are fixed (geostatistics or point-referenced data) or random (point process). Note that in both cases the underlying spatial process will be defined at every point of the study region and not at a discrete number of locations (as in the case of lattice data). A general description of model-based geostatistics can be found in Diggle and Ribeiro Jr. (2007). Point pattern analysis is thoroughly described in Illian et al. (2008), Diggle (2013) and Baddeley et al. (2015), for example.

Modeling and analyzing data observed at specific locations is the primary object of this book and these two cases will be considered in more detail. Next, Gaussian random fields to model continuous spatial processes will be introduced.

2.1.2 The Gaussian random field

To introduce some notation, let $\mathbf{s}$ be any location in a study area $\mathbf{D}$ and let $U(\mathbf{s})$ be the random (spatial) effect at that location. $U(\mathbf{s})$ is a stochastic process, with $\mathbf{s} \in \mathbf{D}$, where $\mathbf{D} \subset \mathbb{R}^d$. Suppose, for example, that $\mathbf{D}$ is a country and data have been measured at geographical locations, over $d = 2$ dimensions within this country.

We denote by $u(\mathbf{s}_i)$, $i = 1, 2, \ldots, n$ a realization of $U(\mathbf{s})$ at n locations. It is commonly assumed that $u(\mathbf{s})$ has a multivariate Gaussian distribution. If $U(\mathbf{s})$ is assumed to be continuous over space, then it is a continuously-indexed Gaussian field (GF). This implies that it is possible to collect these data at any finite set of locations within the study region. To complete the specification of the distribution of $u(\mathbf{s})$, it is necessary to define its mean and covariance.

A very simple option is to define a correlation function based only on the Euclidean distance between locations. This assumes that given two pairs of points separated by the same distance h, they will have the same degree of correlation; Abrahamsen (1997) presents Gaussian random fields and correlation functions.

Now suppose that data y_i have been observed at locations $\mathbf{s}_i$, $i = 1, ..., n$. If an underlying GF generated these data, the parameters of this process can be fitted by considering $y(\mathbf{s}_i) = u(\mathbf{s}_i)$, where observation $y(\mathbf{s}_i)$ is assumed to be a realization of the GF at the location $\mathbf{s}_i$. If it is further assumed that $y(\mathbf{s}_i) = \mu + u(\mathbf{s}_i)$, then there is one more parameter to estimate. It is worth mentioning that the distribution of $u(\mathbf{s})$ at a finite number of points is considered a realization of a multivariate Gaussian distribution. In this case, the likelihood function is the multivariate Gaussian distribution with covariance Σ.

In many situations it is assumed that there is an underlying GF that cannot be directly observed. Instead, observations are data with a measurement error e_i, i.e.,

$$y(\mathbf{s}_i) = u(\mathbf{s}_i) + e_i. \tag{2.1}$$

It is common to assume that e_i is independent of e_j for all $i \neq j$ and e_i follows a Gaussian distribution with zero mean and variance σ_e^2. This additional parameter, σ_e^2, is also known as the "nugget effect" in geostatistics. The covariance of the marginal distribution of $y(\mathbf{s})$ at a finite number of locations is $\Sigma_y = \Sigma + \sigma_e^2 \mathbf{I}$. This is a short extension of the basic GF model, and gives one additional parameter to estimate. For more details about this geostatistics model see, for example, Diggle and Ribeiro Jr. (2007).

The spatial process $u(\mathbf{s})$ if often assumed to be stationary and isotropic. A spatial process is stationary if its statistical properties are invariant under translation; i.e., they are the same at any point of the study region. Isotropy means that the process is invariant under rotation; i.e., their properties do not change regardless of the direction we move around the study region. Stationary and isotropic spatial process play an important role in spatial statistics, as they have desirable properties, such as a constant mean and the fact that the covariance between any two points only depends on their distance and not on their relative position. See, for example, Cressie and Wikle (2011), Section 2.2, for more details about these properties.

The usual way to evaluate the likelihood function, which is just a multivariate Gaussian density for the model in Equation (2.1), usually considers a Cholesky factorization of the covariance matrix (see, for example, Rue and Held, 2005). Because this matrix is dense, this is an operation of order $O(n^3)$, so this is a "big n problem". Some software for geostatistical analysis uses an empirical variogram to fit the parameters of the correlation function. However, this option does not make any assumptions about a likelihood function for the data or uses a multivariate Gaussian distribution for the spatially structured random effect. A good description of these techniques is available in Cressie (1993).

To model spatial dependence with non-Gaussian data, it is usual to assume a likelihood for the data conditional on an unobserved random effect, which is a GF. Such spatial mixed effects models fall under the model-based geostatistics approach (Diggle and Ribeiro Jr., 2007). It is possible to describe the model in Equation (2.1) within a larger class of models: hierarchical models. Suppose that observations y_i have been obtained at locations $\mathbf{s}_i$, $i = 1, ..., n$. The starting model is

$$
\begin{array}{rcl}
y_i | \beta, u_i, \mathbf{F}_i, \phi & \sim & f(y_i | \mu_i, \phi) \\
\mathbf{u} | \theta & \sim & GF(0, \Sigma)
\end{array}
\tag{2.2}
$$

where $\mu_i = h(\mathbf{F}_i^T \beta + u_i)$. Here, $\mathbf{F}_i$ is a matrix of covariates with associated coefficients β, $\mathbf{u}$ is a vector of random effects and $h(\cdot)$ is a function mapping the linear predictor $\mathbf{F}_i^T \beta + u_i$ to $\mathrm{E}(y_i) = \mu_i$. In addition, θ are parameters for the random effect and ϕ is a dispersion parameter of the distribution of the data $f(\cdot)$, which is assumed to be in the exponential family.

To write the GF (with the Gaussian noise as a nugget effect), y_i is assumed to have a Gaussian distribution (with variance σ_e^2), $\mathbf{F}_i^T \beta$ is replaced by β_0 and $\mathbf{u}$ is modeled as a GF. Furthermore, it is possible to consider a multivariate Gaussian distribution for the random effect but it is seldom practical to use the covariance directly for model-based inference. This is shown in Equation (2.3).

$$
\begin{array}{rcl}
y_i | \mu_i, \sigma_e & \sim & N(y_i | \mu_i, \sigma_e^2) \\
\mu_i & = & \beta_0 + u_i \\
\mathbf{u} | \theta & \sim & GF(0, \Sigma)
\end{array}
\tag{2.3}
$$

As mentioned in Section 2.1.1, in the analysis of areal data there are models specified by conditional distributions that imply a joint distribution with a sparse precision matrix. These models are called Gaussian Markov random fields (GMRF) and a good reference is Rue and Held (2005). It is computationally easier to make Bayesian inference when a GMRF is used than when a GF is used, because the computational cost of working with a sparse precision matrix in GMRF models is (typically) $O(n^{3/2})$ in $\mathbb{R}^2$. This makes it easier to conduct analyses with big n.

This basic hierarchical model can be extended in many ways, and some extensions will be considered later. When the general properties of the GF are known, all the practical models that contain, or are based on this random effect can be studied.

2.1.3 The Matérn covariance

A very popular correlation function is the Matérn correlation function. It has a scale parameter $\kappa > 0$ and a smoothness parameter $\nu > 0$. For two locations $\mathbf{s}_i$ and $\mathbf{s}_j$, the stationary and isotropic Matérn correlation function is:

$$Cor_M(U(\mathbf{s}_i), U(\mathbf{s}_j)) = \frac{2^{1-\nu}}{\Gamma(\nu)} (\kappa \parallel \mathbf{s}_i - \mathbf{s}_j \parallel)^{\nu} K_{\nu}(\kappa \parallel \mathbf{s}_i - \mathbf{s}_j \parallel)$$

where $\parallel . \parallel$ denotes the Euclidean distance and K_{ν} is the modified Bessel function of the second kind. The Matérn covariance function is $\sigma_u^2 Cor_M(U(\mathbf{s}_i), U(\mathbf{s}_j))$, where σ_u^2 is the marginal variance of the process.

If $u(\mathbf{s})$ is a realization from $U(\mathbf{s})$ at n locations, $\mathbf{s}_1, ..., \mathbf{s}_n$, its joint covariance matrix can be easily defined as each entry of this joint covariance matrix Σ is $\Sigma_{i,j} = \sigma_u^2 Cor_M(u(\mathbf{s}_i), u(\mathbf{s}_j))$. It is common to assume that $U(.)$ has a zero mean. Hence, we have now completely defined a multivariate distribution for $u(\mathbf{s})$.

To gain a better understanding about the Matérn correlation, samples can be drawn from the GF process. A sample $\mathbf{u}$ is drawn considering $\mathbf{u} = \mathbf{R}^{\top} \mathbf{z}$ where $\mathbf{R}$ is the Cholesky decomposition of the covariance at n locations (see below); i.e., the covariance matrix is equal to $\mathbf{R}^{\top} \mathbf{R}$ with $\mathbf{R}$ an upper triangular matrix, and $\mathbf{z}$ is a vector with n samples drawn from a standard Gaussian distribution. It implies that $E(\mathbf{u}) = E(\mathbf{R}^{\top} \mathbf{z}) = \mathbf{R}^{\top} E(\mathbf{z}) = \mathbf{0}$ and $Var(\mathbf{u}) = \mathbf{R}^{\top} Var(\mathbf{z}) \mathbf{R} = \mathbf{R}^{\top} \mathbf{R}$.

Two useful functions to sample from a GF are defined below:

```
# Matern correlation
cMatern <- function(h, nu, kappa) {
  ifelse(h > 0, besselK(h * kappa, nu) * (h * kappa)^nu /
    (gamma(nu) * 2^(nu - 1)), 1)
}


# Function to sample from zero mean multivariate normal
rmvnorm0 <- function(n, cov, R = NULL) {
  if (is.null(R))
    R <- chol(cov)

  return(crossprod(R, matrix(rnorm(n * ncol(R)), ncol(R))))
}
```

Function `cMatern()` computes the Matérn covariance of two points at distance h, which requires additional parameters `nu` and `kappa`. Function `rmvnorm0()` computes `n` samples from a multivariate Gaussian distribution using the Cholesky decomposition, which needs the covariance matrix `cov` or the upper triangles matrix `R` from a Cholesky decomposition.

Hence, given n locations, function `cMatern()` can be used to compute the Matérn covariance, and function `rmvnorm0()` can be used to draw samples. These steps are collected into the function `book.rMatern`, which is available in the file `spde-book-functions.R`. This file is available from the website of this book.

In order to simplify the visualization of the properties of continuous spatial processes with a Matérn covariance, a set of $n = 249$ locations in the one-dimensional space from 0 to 25 will be considered.

```
# Define locations and distance matrix
loc <- 0:249 / 25
mdist <- as.matrix(dist(loc))
```

Four values for the smoothness parameter ν will be considered. The values for the κ parameter were determined from the range expression $\sqrt{8\nu}/\kappa$, which is the distance that gives correlation near 0.14. By combining the four values for ν with two range values, there are eight parameter configurations.

The values of the different parameters ν, range and κ are created as follows:

```
# Define parameters
nu <- c(0.5, 1.5, 2.5, 5.5)
range <- c(1, 4)
```

Next, the different combinations of the parameters are put together in a `matrix`:

```
# Covariance parameter scenarios
params <- cbind(nu = rep(nu, length(range)),
  range = rep(range, each = length(nu)))
```

Sampled values depend on the covariance matrix and the noise considered for the standard Gaussian distribution. In the following example, five vectors of size n are drawn from the standard Gaussian distribution. These five standard Gaussian observations, `z` in the code below, were the same across the eight parameter configurations in order to keep track of what the different parameter configurations are doing.

```
# Sample error
set.seed(123)
z <- matrix(rnorm(nrow(mdist) * 5), ncol = 5)
```

Therefore, there is a set of 40 different realizations, five for each parameter configuration:

```
# Compute the correlated samples
# Scenarios (i.e., different set of parameters)
yy <- lapply(1:nrow(params), function(j) {
  param <- c(params[j, 1], sqrt(8 * params[j, 1]) / params[j, 2],
    params[j, 2])
  v <- cMatern(mdist, param[1], param[2])

  # fix the diagonal to avoid numerical issues
  diag(v) <- 1 + 1e-9

  # Parameter scenario and computed sample
  return(list(params = param, y = crossprod(chol(v), z)))
})
```

These samples are shown in the eight plots in Figure 2.2. One important point
to observe here is the main spatial trend in the samples, as it seems not to
depend on the smoothness parameter ν. In order to appreciate the main trend,
consider one of the five samples (i.e., one of the colors). Then, compare the
curves for different values of the smoothness parameter. If noise is added to
the smoothest curve, and the resulting curve is compared to the other curves
fitted using different smoothness parameters, it is difficult to disentangle what
is due to noise from what is due to smoothness. Therefore, in practice, the
smoothness parameter is usually fixed and a noise term is added.

2.1.4   Simulation of a toy data set

Now, a sample from the model in Equation (2.1) will be drawn and it will
be used in Section 2.3. A set of $n = 100$ locations in the unit square will be
considered. This sample will have a higher density of locations in the bottom
left corner than in the top right corner. The R code to do this is:

```
n <- 200
set.seed(123)
pts <- cbind(s1 = sample(1:n / n - 0.5 / n)^2,
  s2 = sample(1:n / n - 0.5 / n)^2)
```

To get a (lower triangular) matrix of distances, the `dist()` function can be
used as follows:

```
dmat <- as.matrix(dist(pts))
```

The chosen parameters for the Matérn covariance are $\sigma_u^2 = 5$, $\kappa = 7$ and $\nu = 1$.

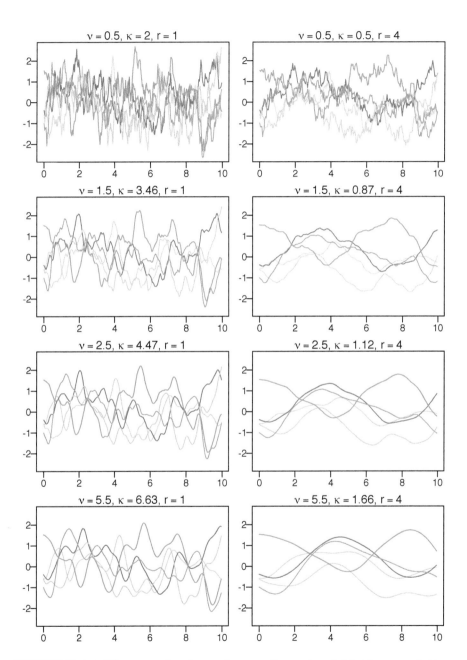

FIGURE 2.2 Five samples from the one-dimensional Matérn correlation function for two different range values (each column of plots) and four different values for the smoothness parameter (each line of plots).

The mean is set to be $\beta_0 = 10$ and the nugget parameter is $\sigma_e^2 = 0.3$. These values are declared using:

```
beta0 <- 10
sigma2e <- 0.3
sigma2u <- 5
kappa <- 7
nu <- 1
```

Now, a sample is obtained from a multivariate distribution with constant mean equal to β_0 and covariance $\sigma_e^2 \mathbf{I} + \Sigma$, which is the marginal covariance of the observations. Σ is the Matérn covariance of the spatial process, `mcor` in the code below.

```
mcor <- cMatern(dmat, nu, kappa)
mcov <- sigma2e * diag(nrow(mcor)) + sigma2u * mcor
```

Next, the sample is obtained by considering the Cholesky factor times a unit variance noise and add the mean:

```
R <- chol(mcov)
set.seed(234)
y1 <- beta0 + drop(crossprod(R, rnorm(n)))
```

Figure 2.3 shows these simulated data in a graph of the locations where the size of the points is proportional to the simulated values.

These data will be used as a toy example in this tutorial. It is available in the R-INLA package and can be loaded by typing:

```
data(SPDEtoy)
```

2.2   The SPDE approach

Rue and Tjelmeland (2002) propose to approximate a continuous Gaussian field using GMRFs, and demonstrated good fits for a range of commonly used covariance functions. Although these "proof-of-concept" results were interesting, their approach was less practical. All fits had to be precomputed on a regular grid only. The real breakthrough, came with Lindgren et al. (2011), who considered a stochastic partial differential equation (SPDE) whose

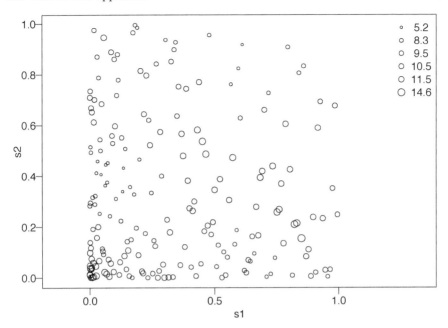

FIGURE 2.3 The simulated toy example data.

solution is a GF with Matérn correlation. Lindgren et al. (2011) propose a new approach to represent a GF with Matérn covariance, as a GMRF, by representing a solution of the SPDE using the finite element method. This was possible only for some values of the smoothness ν, where the continuous indexed random field was Markov (Rozanov (1977)). The benefit is that the GMRF representation of the GF, which can be computed explicitly, provides a sparse representation of the spatial effect through a sparse precision matrix. This enables the nice computational properties of the GMRFs which can then be implemented in the `INLA` package.

Warning. In this section the main results in Lindgren et al. (2011) are summarized. If your purpose does not include understanding the underlying methodology in depth, you can skip this section. If you keep reading this section and find it difficult, do not be discouraged. You will still be able to use `INLA` for applications even if you have only a limited grasp of what is "under the hood".

In this section we have tried to provide an intuitive approach to SPDE. However, if you are interested in the full details, they are in the Appendix of Lindgren et al. (2011). In a few words, it uses the Finite Element Method (FEM, Ciarlet, 1978; Brenner and Scott, 2007; Quarteroni and Valli, 2008) to provide a solution to a SPDE. They develop this solution by considering basis functions carefully chosen to preserve the sparse structure of the resulting precision matrix for the random field at a set of mesh nodes. This provides an explicit link between a continuous random field and a GMRF representation, which allows efficient computations.

2.2.1   First result

The first main result provided in Lindgren et al. (2011) is that a GF with a generalized covariance function, obtained when $\nu > 0$ in the Matérn correlation function is a solution to a SPDE. This extends the result obtained by Besag (1981). A more statistical way to look at this result is by considering a regular two-dimensional lattice with number of sites tending to infinity. The representation of sites in a lattice can be seen in Figure 2.4.

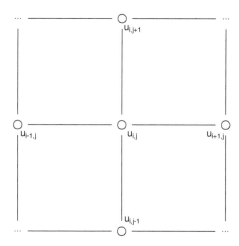

FIGURE 2.4 Representation of sites in a two-dimensional lattice to estimate a spatial process.

In this case the full conditional distribution at the site i, j has expectation

$$E(u_{i,j}|u_{-i,j}) = \frac{1}{a}(u_{i-1,j} + u_{i+1,j} + u_{i,j-1} + u_{i,j+1})$$

and variance $Var(u_{i,j}|u_{-i,j}) = 1/a$, with $|a| > 4$. In the representation using a precision matrix, for a single site, only the upper right quadrant is shown and with a as the central element, such that

$$\begin{vmatrix} -1 & \\ a & -1 \end{vmatrix} \qquad (2.4)$$

A GF $U(\mathbf{s})$ with Matérn covariance is a solution to the following linear fractional SPDE:

$$(\kappa^2 - \Delta)^{\alpha/2} u(\mathbf{s}) = \mathbf{W}(\mathbf{s}), \quad \mathbf{s} \in \mathbb{R}^d, \quad \alpha = \nu + d/2, \quad \kappa > 0, \quad \nu > 0.$$

Here, Δ is the Laplacian operator and $\mathbf{W}(\mathbf{s})$ denotes a spatial white noise Gaussian stochastic process with unit variance.

Lindgren et al. (2011) show that, for $\nu = 1$ and $\nu = 2$, the GMRF representation is a convolution of processes with precision matrix as in Equation (2.4). The resulting upper right quadrant precision matrix and center can be expressed as a convolution of the coefficients in Equation (2.4). For $\nu = 1$, using this representation, we have:

$$
\begin{vmatrix}
1 & & \\
-2a & 2 & \\
4 + a^2 & -2a & 1
\end{vmatrix}
\tag{2.5}
$$

and, for $\nu = 2$:

$$
\begin{vmatrix}
-1 & & & \\
3a & -3 & & \\
-3(a^2 + 3) & 6a & -3 & \\
a(a^2 + 12) & -3(a^2 + 3) & 3a & -1
\end{vmatrix}
\tag{2.6}
$$

An intuitive interpretation of this result is that as the smoothness parameter ν increases, the precision matrix in the GMRF representation becomes less sparse. Greater density of the matrix is due to the fact that the conditional distributions depend on a wider neighborhood. Notice that it does not imply that the conditional mean is an average over a wider neighborhood.

Conceptually, this is like going from a first order random walk to a second order one. To understand this point, let us consider the precision matrix for the first order random walk, its square, and the precision matrix for the second order random walk.

```
q1 <- INLA:::inla.rw1(n = 5)
q1
## 5 x 5 sparse Matrix of class "dgTMatrix"
##
## [1,]  1 -1  .  .  .
## [2,] -1  2 -1  .  .
## [3,]  . -1  2 -1  .
## [4,]  .  . -1  2 -1
## [5,]  .  .  . -1  1
# Same inner pattern as for RW2
crossprod(q1)
## 5 x 5 sparse Matrix of class "dsCMatrix"
```

```
##
## [1,]   2 -3   1   .   .
## [2,]  -3  6  -4   1   .
## [3,]   1 -4   6  -4   1
## [4,]   .  1  -4   6  -3
## [5,]   .   .   1  -3   2
INLA:::inla.rw2(n = 5)
## 5 x 5 sparse Matrix of class "dgTMatrix"
##
## [1,]   1 -2   1   .   .
## [2,]  -2  5  -4   1   .
## [3,]   1 -4   6  -4   1
## [4,]   .  1  -4   5  -2
## [5,]   .   .   1  -2   1
```

As can be seen in the previous matrices, the differences between the precision matrix of a random walk of order 2 and the crossproduct of a precision matrix of a random walk of order 1 appear at the upper-left and bottom-right corners. The precision matrix for $\alpha = 2$, $\mathbf{Q}_2 = \mathbf{Q}_1 \mathbf{C}^{-1} \mathbf{Q}_1$, is a standardized square of the precision matrix for $\alpha = 1$, $\mathbf{Q}_1$. Matrices $\mathbf{Q}_1$, $\mathbf{C}$ and $\mathbf{Q}_2$ are fully described later in Section 2.2.2.

2.2.2   Second result

Point data are seldom located on a regular grid but distributed irregularly. Lindgren et al. (2011) provide a second set of results that give a solution for the case of irregular grids. This was made considering the FEM, which is widely used in engineering and applied mathematics to solve differential equations.

The domain can be divided into a set of non-intersecting triangles, which may be irregular, where any two triangles meet in at most a common edge or corner. The three corners of a triangle are named vertices or nodes. The solution for the SPDE and its properties will depend on the basis functions used. Lindgren et al. (2011) choose basis functions carefully in order to preserve the sparse structure of the resulting precision matrix.

The approximation is:

$$u(\mathbf{s}) = \sum_{k=1}^{m} \psi_k(\mathbf{s}) w_k \tag{2.7}$$

where ψ_k are basis functions and w_k are Gaussian distributed weights, $k = 1, ..., m$, with m the number of vertices in the triangulation. A stochastic weak solution was considered to show that the joint distribution for the weights determines the full distribution in the continuous domain.

More information about the weak solution and how to use the FEM in this case can be found in Bakka (2018).

We will now illustrate how the process $u(\mathbf{s})$ can be approximated to any point inside the triangulated domain. First, we consider the one-dimensional case in Figure 2.5. We have a set of six piece-wise linear basis functions shown at the top of this figure. The value of each of such basis functions is one at the basis knot, decreases linearly to zero until the center of the next basis function and is zero elsewhere. Thus for any point between two basis knots only two basis functions have a non-zero value. At the bottom of Figure 2.5, we have the sine function that was approximated using these basis functions. Note that we have used irregularly spaced knots here. Thus, the intervals between the knots do not need to be equally spaced (in a similar way as triangles in a mesh do not need to be equal). For example, it will be better to place more knots (or triangles in space) where the function changes more rapidly, as we did considering a knot at 1.5, shown in Figure 2.5. From this figure it is also clear that it would be better to have one knot around 4.5 as well.

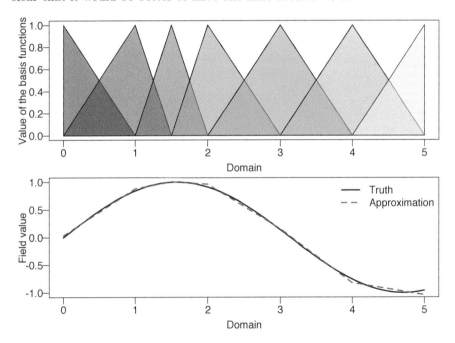

FIGURE 2.5 One dimensional approximation illustration. The one dimensional piece-wise linear basis functions (top). A function and its approximation (bottom).

We now illustrate the approximation in two dimensional space considering piece-wise linear basis functions. These are based on triangles as a generalization of the idea in one dimension. Consider the big triangle shown on the top left of Figure 2.6, the point shown as a red dot inside this big triangle, and the three small triangles formed by this point and each vertex of the big one. The numbers at the vertex of the big triangle are equal to the area of the opposite triangle inside the big one (not formed by this vertex), divided by the area total of the big triangle. Thus, the three numbers sum up to one. These three numbers are the value of the basis function centered at the vertices of the big triangle being evaluated at the red point. These three values are considered for the approximation as they are the coefficients that multiply the function at each vertex of the big triangle.

Writing in matrix form, we have the projector matrix $\mathbf{A}$ and when a point is inside a triangle, there are three non-zero values in the corresponding line of $\mathbf{A}$. When the point is along an edge, there are two non-zero values and when the point is on top of a triangle vertex there is only one non-zero (which is equal to one). This is just an illustration of the barycentric coordinates of the point with respect to the coordinates of the triangle vertices known also as areal coordinates for this particular case.

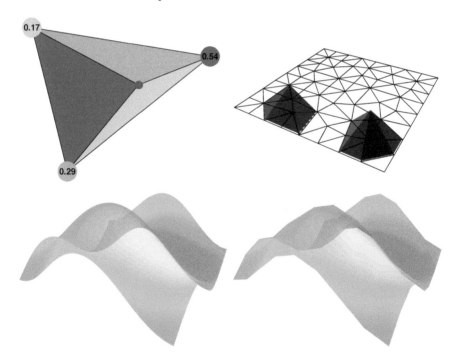

FIGURE 2.6 Two dimensional approximation illustration. A triangle and the areal coordinates for the point in red (top left). All the triangles and the basis function for two of them (top right). A true field for illustration (bottom left) and its approximated version (bottom right).

We will now focus on the resulting precision matrix of the observations $\mathbf{Q}_{\alpha,\kappa}$.

It does consider the triangulation and the basis functions. The precision matrix matches the first result when applied to a regular grid. Consider the set of $m \times m$ matrices $\mathbf{C}$, $\mathbf{G}$ and $\mathbf{K}_\kappa$ with entries

$$\mathbf{C}_{i,j} = \langle \psi_i, \psi_j \rangle, \qquad \mathbf{G}_{i,j} = \langle \nabla \psi_i, \nabla \psi_j \rangle, \qquad (\mathbf{K}_\kappa)_{i,j} = \kappa^2 C_{i,j} + G_{i,j} \,.$$

Here, $\langle \cdot, \cdot \rangle$ denotes the inner product and ∇ the gradient.

The precision matrix $\mathbf{Q}_{\alpha,\kappa}$ as a function of κ^2 and α can be written as:

$$\begin{array}{rcl} \mathbf{Q}_{1,\kappa} & = & \mathbf{K}_\kappa = \kappa^2 \mathbf{C} + \mathbf{G}, \\ \mathbf{Q}_{2,\kappa} & = & \mathbf{K}_\kappa \mathbf{C}^{-1} \mathbf{K}_\kappa = \kappa^4 \mathbf{C} + 2\kappa^2 \mathbf{G} + \mathbf{G}\mathbf{C}^{-1}\mathbf{G} \\ \mathbf{Q}_{\alpha,\kappa} & = & \mathbf{K}_\kappa \mathbf{C}^{-1} \mathbf{Q}_{\alpha-2,\kappa} \mathbf{C}^{-1} \mathbf{K}_\kappa, \quad \text{for } \alpha = 3, 4, \dots . \end{array} \qquad (2.8)$$

As the matrix $\mathbf{C}$ is dense, it can be replaced by a diagonal matrix $\tilde{\mathbf{C}}$ with

$$\tilde{\mathbf{C}}_{i,i} = \langle \psi_i, 1 \rangle, \qquad (2.9)$$

which is common when working with FEM. In this case, since $\tilde{\mathbf{C}}$ is diagonal, $\mathbf{K}_\kappa$ is as sparse as $\mathbf{G}$.

As an example, we consider a set of seven points and build a mesh around four of them with an adequate choice of the arguments in order to have a didactic illustration of the FEM. The dual mesh is also built. Finally, the code extracts matrices $\mathbf{C}$, $\mathbf{G}$ and $\mathbf{A}$:

```
# This 's' factor will only change C, not G
s <- 3
pts <- cbind(c(0.1, 0.9, 1.5, 2, 2.3, 2, 1),
  c(1, 1, 0, 0, 1.2, 1.9, 2)) * s
n <- nrow(pts)
mesh <- inla.mesh.2d(pts[-c(3, 5, 7), ], max.edge = s * 1.7,
  offset = s / 4, cutoff = s / 2, n = 6)
m <- mesh$n
dmesh <- book.mesh.dual(mesh)
fem <- inla.mesh.fem(mesh, order = 1)
A <- inla.spde.make.A(mesh, pts)
```

Here, function `book.mesh.dual()` is used to create the dual mesh polygons (see Figure 2.7). Furthermore, function `inla.mesh.fem()` is used to create matrices $\mathbf{C}$ and $\mathbf{G}$, while function `inla.spde.make.A()` creates the projector matrix $\mathbf{A}$.

We can gain intuition about this result by considering the structure of each matrix, which is detailed in Appendix A.2 in Lindgren et al. (2011). It may be easier to understand it by considering the plots in Figure 2.7. In this figure, a mesh with 8 nodes is shown in thicker border lines. The corresponding dual mesh forms a collection of polygons around each vertex of the mesh.

The dual mesh is a set of polygons, one polygon for each vertex. Each polygon is formed by the mid of each of the edges that connects to the vertex, and the centroids of the triangles that the vertex is a corner of. Notice that for the vertices at the boundary of the mesh, this vertex is one point of the dual polygon as well. The area of each dual polygon is equal to $\tilde{\mathbf{C}}_{ii}$.

Equivalently, $\tilde{\mathbf{C}}_{ii}$ is equal to the sum of one third the area of each triangle that the vertex i is part of. Notice that each polygon around each mesh node is formed by one third of the triangles that it is part of. The $\tilde{\mathbf{C}}$ matrix is diagonal.

Matrix $\mathbf{G}$ reflects the connectivity of the mesh nodes. Nodes not connected by edges have the corresponding entry as zero. Values do not depend on the size of the triangles as they are scaled by the area of the triangles. For more detailed information, see Appendix 2 in Lindgren et al. (2011).

As stated above, the resulting precision matrix for increasing ν is a convolution of the precision matrix for $\nu - 1$ with a scaled $\mathbf{K}_\kappa$. It still implies a denser precision matrix when working with $\kappa\mathbf{C} + \mathbf{G}$.

The $\mathbf{Q}$ precision matrix can be generalized to fractional values of α (or ν) using a Taylor approximation. See the author's discussion response in Lindgren et al. (2011). This approximation leads to a polynomial of order $p = \lceil\alpha\rceil$ for the precision matrix:

$$\mathbf{Q} = \sum_{i=0}^{p} b_i \mathbf{C}(\mathbf{C}^{-1}\mathbf{G})^i. \tag{2.10}$$

For $\alpha = 1$ and $\alpha = 2$, the precision matrix is the same as in Equation (2.8). For $\alpha = 1$, the values of the coefficients of the Taylor approximation are $b_0 = \kappa^2$ and $b_1 = 1$. For the case $\alpha = 2$, the coefficients are $b_0 = \kappa^4$, $b_1 = \alpha\kappa^4$ and $b_2 = 1$.

For fractional $\alpha = 1/2$, it holds that $b_0 = 3\kappa/4$ and $b_1 = \kappa^{-1}3/8$. And for $\alpha = 3/2$ (and $\nu = 0.5$, the exponential case), the values are $b_0 = 15\kappa^3/16$, $b_1 = 15\kappa/8$ and $b_2 = 15\kappa^{-1}/128$. Using these results combined with a recursive construction, for $\alpha > 2$, GMRF approximations can be obtained for all positive integers and half-integers.

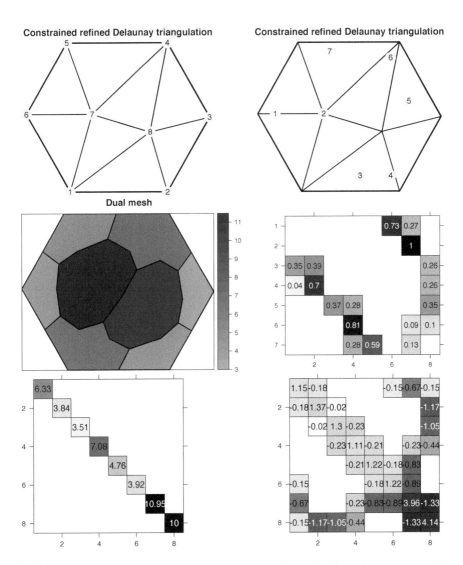

FIGURE 2.7 A mesh and its nodes numbered (top left) and the mesh with some points numbered (top right). The dual mesh polygons (mid left) and **A** matrix (mid right). The associated **C** matrix (bottom left) and **G** matrix (bottom center).

2.3 A toy example

In this example we will fit a simple geostatistical model to the toy dataset
simulated in Section 2.1.4. This dataset is also available in the INLA package
and it can be loaded as follows:

```
data(SPDEtoy)
```

This dataset is a three-column data.frame. Coordinates are in the first two
columns and the response in the third column.

```
str(SPDEtoy)
## 'data.frame':    200 obs. of  3 variables:
## $ s1: num  0.0827 0.6123 0.162 0.7526 0.851 ...
## $ s2: num  0.0564 0.9168 0.357 0.2576 0.1541 ...
## $ y : num  11.52 5.28 6.9 13.18 14.6 ...
```

2.3.1 SPDE model definition

Given n observations y_i, $i = 1, ..., n$, at locations $\mathbf{s}_i$, the following model can
be defined:

$$
\begin{aligned}
\mathbf{y}|\beta_0, \mathbf{u}, \sigma_e^2 &\sim N(\beta_0 + \mathbf{Au}, \sigma_e^2) \\
\mathbf{u} &\sim GF(0, \Sigma)
\end{aligned}
$$

where β_0 is the intercept, $\mathbf{A}$ is the projector matrix and $\mathbf{u}$ is a spatial Gaussian
random field. Note that the projector matrix $\mathbf{A}$ links the spatial Gaussian
random field (defined using the mesh nodes) to the locations of the observed
data.

This mesh must cover the entire spatial domain of interest. More details on
mesh building are given in Section 2.6. Here, the fifth mesh built in Section
2.6.3 will be used. In the following R code, a domain is first defined to create
the mesh:

```
pl.dom <- cbind(c(0, 1, 1, 0.7, 0), c(0, 0, 0.7, 1, 1))
mesh5 <- inla.mesh.2d(loc.domain = pl.dom, max.e = c(0.092, 0.2))
```

The SPDE model in the original parameterization can be built using function
inla.spde2.matern(). The main arguments of this function are the mesh
object (mesh) and the α parameter (alpha), which is related to the smoothness
parameter of the process. The parameterization is flexible and can be defined

by the user, with a default choice controlling the log of τ and κ, that jointly control the variance and correlation range.

Instead of using the default parameterization, it is more intuitive to control the parameters through the marginal standard deviation and the range, $\sqrt{8\nu}/\kappa$. For details on this parameterization see Lindgren (2012). In addition, when defining the SPDE model a set of priors for both parameters is also required. The `inla.spde2.pcmatern()` uses this parameterization to set the Penalized Complexity prior, PC-prior, as derived in Fuglstad et al. (2018). The Penalized Complexity prior is under the framework reviewed in Section 1.6.5.

The PC-prior was derived for the practical range, or just range, which is the distance such that the correlation is around 0.139, and the marginal standard deviation of the field, σ. The way of setting these priors for σ is that we do need to set σ_0 and p such that $P(\sigma > \sigma_0) = p$. In our example we will set $\sigma_0 = 10$ and $p = 0.01$ and these values are passed to the `inla.spde2.matern()` function as a vector in the next code. For the practical range parameter the setting is that we have to choose r_0 and p such that $P(r < r_0) = p$. We have to account for the fact that in our example the domain is the $[0, 1] \times [0, 1]$ square. We can set the PC-prior for the median by setting $p = 0.5$ and in the next code we consider a prior median equal to 0.3.

The smoothness parameter ν is fixed as $\alpha = \nu + d/2 \in [1, 2]$. Next, we build the SPDE model considering that the toy data was simulated with $\alpha = 2$, which is actually the default value in the `inla.spde2.pcmatern()` function:

```
spde5 <- inla.spde2.pcmatern(
  # Mesh and smoothness parameter
  mesh = mesh5, alpha = 2,
  # P(practic.range < 0.3) = 0.5
  prior.range = c(0.3, 0.5),
  # P(sigma > 1) = 0.01
  prior.sigma = c(10, 0.01))
```

2.3.2 Projector matrix

The second step when setting an SPDE model is to build a projector matrix. The projector matrix contains the basis function value for each basis, with one basis function at each column. It will be used to interpolate the random field that is being modeled at the mesh nodes. For details, see Section 2.2.2. The projector matrix can be built with the `inla.spde.make.A()` function. Considering the basis function at each mesh vertex, the basis function value for one point within one triangle is computed as illustrated in Figure 2.6. Thus, the value for the random field is the projection of a plane (formed by the random field value at these three mesh points) to this point location. This is why we call a projector matrix to the matrix that stores in each line a different basis function evaluated at a location point. This matrix is sparse since no more than three elements in each line are non-zero. Also, the sum of each row is equal to one.

Using the toy dataset and example mesh **mesh5**, the projector matrix can be computed as follows:

```
coords <- as.matrix(SPDEtoy[, 1:2])
A5 <- inla.spde.make.A(mesh5, loc = coords)
```

This matrix has dimension equal to the number of data locations times the number of vertices in the mesh:

```
dim(A5)
## [1] 200 489
```

Because each point location is inside one of the triangles there are exactly three non-zero elements on each line:

```
table(rowSums(A5 > 0))
##
##    3
## 200
```

Furthermore, these three elements on each line sum up to one:

```
table(rowSums(A5))
##
##    1
## 200
```

The reason why they sum up to one is that each matrix element is a basis function evaluated at a point location, and the basis functions sum up to one at each location. Multiplication of this matrix by a vector representing a continuous function evaluated at the locations gives the interpolation of this function at the point locations.

There are some columns in the projector matrix all of whose elements equal zero:

```
table(colSums(A5) > 0)
##
```

```
## FALSE   TRUE
##   237   252
```

These columns correspond to triangles with no point location inside. These columns can be dropped. The `inla.stack()` function (Section 2.3.3) does this automatically.

When there is a mesh where every point location is at a mesh vertex, each line on the projector matrix has exactly one non-zero element. This is the case for the `mesh1` built in Section 2.6.3:

```
A1 <- inla.spde.make.A(mesh1, loc = coords)
```

In this case, all the non-zero elements in the projector matrix are equal to one:

```
table(as.numeric(A1))
##
##       0       1
## 580400     200
```

Each element is equal to one in this case because the location points are actually one of the mesh nodes and thus the weight of the basis function at one mesh node is equal to one.

2.3.3   The data stack

The `inla.stack()` function is useful for organizing data, covariates, indices and projector matrices, all of which are important when constructing an SPDE model. `inla.stack()` helps to control the way effects are projected in the linear predictor. Detailed examples including one dimensional, replicated random field and space-time models are presented in Lindgren (2012) and Lindgren and Rue (2015), which also details the mathematical theory for the stack method.

In the toy example, there is a linear predictor that can be written as

$$\eta^* = \mathbf{1}\beta_0 + \mathbf{A}\mathbf{u} \,.$$

The first term on the right-hand side represents the intercept, while the second represents the spatial effect. Each term is represented as a product of a projector matrix and an effect.

The solution obtained with the Finite Element Method considered when implementing the SPDE models builds the model on the mesh nodes. Usually, the number of nodes is not equal to the number of locations for which we have observations. The `inla.stack()` function allows us to work with predictors that includes terms with different dimensions. The three main `inla.stack()` arguments are a vector list with the data (`data`), a list of projector matrices (each related to one block effect, `A`) and the list of effects (`effects`). Optionally, a label can be assigned to the data stack (using argument `tag`).

Two projector matrices are needed: the projector matrix for the latent field and a matrix that is a one-to-one map of the covariate and the response. The latter matrix can simply be a constant rather than a diagonal matrix.

The following R code will take the toy example data and use function `inla.stack()` to put all these three elements (i.e., data, projector matrices and effects) together:

```
stk5 <- inla.stack(
  data = list(resp = SPDEtoy$y),
  A = list(A5, 1),
  effects = list(i = 1:spde5$n.spde,
    beta0 = rep(1, nrow(SPDEtoy))),
  tag = 'est')
```

The `inla.stack()` function automatically eliminates any column in a projector matrix that has a zero sum, and it generates a new and simplified matrix. The function `inla.stack.A()` extracts a simplified matrix to use as an observation predictor matrix with the `inla()` function, while the `inla.stack.data()` function extracts the corresponding data.

The simplified projector matrix from the stack consists of the simplified projector matrices, where each column holds one effect block. Hence, its dimensions are:

```
dim(inla.stack.A(stk5))
## [1] 200 253
```

In the toy example, there is one column more than the number of columns with non-zero elements in the projector matrix. This extra column is due to the intercept and all values are equal to one.

2.3.4   Model fitting and some results

To fit the model, the intercept in the formula must be removed and added as a covariate term in the list of effects, so that all the covariate terms in the formula can be included in a projector matrix. Then, the matrix of predictors is passed to the `inla()` function in its `control.predictor` argument, as follows:

```
res5 <- inla(resp ~ 0 + beta0 + f(i, model = spde5),
  data = inla.stack.data(stk5),
  control.predictor = list(A = inla.stack.A(stk5)))
```

The `inla()` function returns an object that is a set of several results. It includes summaries, marginal posterior densities of each parameter in the model, the regression parameters, each element that is a latent field, and all the hyperparameters.

The summary of the intercept β_0 is obtained with the following R code:

```
res5$summary.fixed
##         mean      sd 0.025quant  0.5quant  0.975quant   mode
## beta0 9.473 0.6793      8.053    9.488       10.81 9.511
##              kld
## beta0 4.473e-10
```

Similarly, the summary of the precision of the Gaussian likelihood, i.e., $1/\sigma_e^2$, can be obtained as follows:

```
res5$summary.hyperpar[1, ]
##                                              mean      sd
## Precision for the Gaussian observations 2.948 0.4723
##                                         0.025quant 0.5quant
## Precision for the Gaussian observations    2.118    2.914
##                                         0.975quant   mode
## Precision for the Gaussian observations    3.973 2.852
```

A marginal distribution in the `inla()` output consists of two vectors. One is a set of values on the range of the parameter space with posterior marginal density bigger than zero and another is the posterior marginal density at each one of these values. Any posterior marginal can be transformed. For example, if the posterior marginal for σ_e is required, the square root of σ_e^2, it can be obtained as follows:

```
post.se <- inla.tmarginal(function(x) sqrt(1 / exp(x)),
  res5$internal.marginals.hyperpar[[1]])
```

Now, it is possible to obtain summary statistics from this marginal distribution:

```
inla.emarginal(function(x) x, post.se)
## [1] 0.588
```

```
inla.qmarginal(c(0.025, 0.5, 0.975), post.se)
## [1] 0.5023 0.5857 0.6860
inla.hpdmarginal(0.95, post.se)
##                 low    high
## level:0.95 0.4985 0.6814
inla.pmarginal(c(0.5, 0.7), post.se)
## [1] 0.02168 0.98655
```

In the previous code, function `inla.emarginal()` is used to compute the posterior expectation of a function on the parameter, function `inla.qmarginal()` computes quantiles from the posterior marginal, function `inla.hpdmarginal()` computes a highest posterior density (HPD) interval and function `inla.pmarginal()` can be used to obtain posterior probabilities. Figure 2.8 shows the posterior marginals of the precision and the standard deviation, which has been created with the following code:

```
par(mfrow = c(1, 2), mar = c(3, 3, 1, 1), mgp = c(2, 1, 0))
plot(res5$marginals.hyperpar[[1]], type = "l", xlab = "precision",
     ylab = "density", main = "Precision")
plot(post.se, type = "l", xlab = "st. deviation",
     ylab = "density", main = "Standard deviation")
```

In cases with weakly identifiable parameters, transforming the posterior marginal densities leads to a loss of accuracy. To reduce that loss, one can operate more directly on the marginal densities without first transforming:

```
post.orig <- res5$marginals.hyperpar[[1]]
fun <- function(x) rev(sqrt(1 / x)) # Use rev() to preserve order
ifun <- function(x) rev(1 / x^2)
inla.emarginal(fun, post.orig)
## [1] 0.5279
fun(inla.qmarginal(c(0.025, 0.5, 0.975), post.orig))
## [1] 0.5022 0.5859 0.6868
fun(inla.hpdmarginal(0.95, post.orig))
## [1] 0.5072 0.6963
inla.pmarginal(ifun(c(0.5, 0.7)), post.orig)
## [1] 0.01413 0.97806
```

Note that the HPD interval is not invariant to changes of variables, and if a specific interpretation is required one therefore has to start by transforming the density to that parameterization.

The posterior marginals for the parameters of the latent field are visualized in Figure 2.9 with:

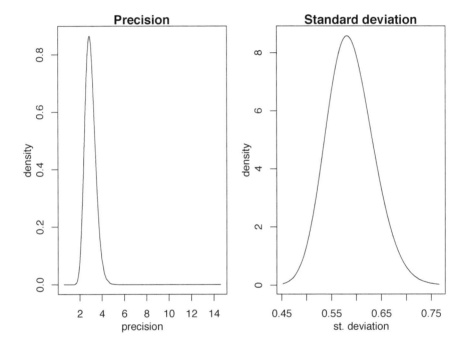

FIGURE 2.8 Posterior marginals of the precision (top) and the standard deviation (bottom).

```
par(mfrow = c(2, 1), mar = c(3, 3, 1, 1), mgp = c(2, 1, 0))
plot(res5$marginals.hyperpar[[2]], type = "l",
  xlab = expression(sigma), ylab = 'Posterior density')
plot(res5$marginals.hyperpar[[3]], type = "l",
  xlab = 'Practical range', ylab = 'Posterior density')
```

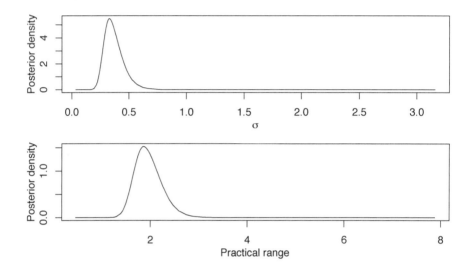

FIGURE 2.9 Posterior marginal distribution for σ (left) and the practical range (right).

Furthermore, summary statistics and HPD intervals can also be computed, and these marginals can be plotted as well to visualize them.

2.4   Projection of the random field

A common objective when dealing with spatial data collected at some locations is the prediction on a fine grid of the spatial model to get high resolution maps. In this section we show how to do this prediction for the random field term only. In Section 2.5, prediction of the outcome will be described. In this example, the random field will be predicted at three target locations: (0.1, 0.1), (0.5, 0.55), (0.7, 0.9). These points are defined in the following code:

```
pts3 <- rbind(c(0.1, 0.1), c(0.5, 0.55), c(0.7, 0.9))
```

The predictor matrix for the target locations is:

```
A5pts3 <- inla.spde.make.A(mesh5, loc = pts3)
dim(A5pts3)
## [1]    3 489
```

In order to visualize only the columns with non-zero elements of this matrix, the following code can be run:

```
jj3 <- which(colSums(A5pts3) > 0)
A5pts3[, jj3]
## 3 x 9 sparse Matrix of class "dgCMatrix"
##
## [1,] .    .    0.054 .    .    .    0.5 0.44 .
## [2,] 0.22 .    .    0.32 .    0.46 .    .    .
## [3,] .    0.12 .    .    0.27 .    .    .    0.61
```

This projector matrix can then be used for interpolating a functional of the random field, for example, the posterior mean. This interpolation is a computationally very cheap operation, since it is a matrix vector product where the matrix is sparse, having only up to three non-zero elements in each row.

```
drop(A5pts3 %*% res5$summary.random$i$mean)
## [1]    0.3114   3.1006 -2.7230
```

2.4.1   Projection on a grid

By default, the `inla.mesh.projector()` function computes a projector matrix automatically for a grid of points over a square that contains the mesh. It can be used to get the map of the random field on a fine grid. To get a projector matrix on a grid in the domain $[0, 1] \times [0, 1]$ these limits can be passed to the `inla.mesh.projector()` function:

```
pgrid0 <- inla.mesh.projector(mesh5, xlim = 0:1, ylim = 0:1,
  dims = c(101, 101))
```

Then, the projection of the posterior mean and the posterior standard deviation can be obtained with the following code:

```
prd0.m <- inla.mesh.project(pgrid0,  res5$summary.random$i$mean)
prd0.s <- inla.mesh.project(pgrid0,  res5$summary.random$i$sd)
```

The values projected on the grid are available in Figure 2.10.

2.5   Prediction

Another quantity of interest when modeling spatially continuous processes is
the prediction of the expected value on a target location for which data have
not been observed. In a similar way as in the previous section, it is possible to
compute the marginal distribution of the expected value at the target location
or to make a projection of some functional of it, specifically, considering that

$$\mathbf{y} \sim N(\mu = \mathbf{1}\beta_0 + \mathbf{A}\mathbf{u}, \sigma_e^2 \mathbf{I}).$$

We will be computing the posterior distribution of μ. This problem is often
called prediction.

2.5.1   Joint estimation and prediction

In this case, we just set a scenario for the prediction and include it in the
stack used in the model fitting. This is similar to the stack created to predict
the random field, but here all the fixed effects have to be considered in the
predictor and effects slots of the stack. In our case we just have the intercept:

```
stk5.pmu <- inla.stack(
  data = list(resp = NA),
  A = list(A5pts3, 1),
  effects = list(i = 1:spde5$n.spde, beta0 = rep(1, 3)),
  tag = 'prd5.mu')
```

This stack is then joined to the data stack to fit the model again. Notice we
can save time by considering the previous fitted model parameters here (by
using parameter `control.mode`):

```
stk5.full <- inla.stack(stk5, stk5.pmu)
r5pmu <- inla(resp ~ 0 + beta0 + f(i, model = spde5),
  data = inla.stack.data(stk5.full),
  control.mode = list(theta = res5$mode$theta, restart = FALSE),
  control.predictor = list(A = inla.stack.A(
    stk5.full), compute = TRUE))
```

The fitted values in an `inla` object are summarized in a single `data.frame` for
all the observations in the dataset. In order to find the predicted values for the
values with missing observations, the index to their rows in the `data.frame`
must be found first. This index can be obtained from the full stack by indicating
the tag in the corresponding stack, as follows:

```
indd3r <- inla.stack.index(stk5.full, 'prd5.mu')$data
indd3r
## [1] 201 202 203
```

To get the summary of the posterior distributions of μ at the target location's the index must be passed to the `data.frame` with the summary statistics:

```
r5pmu$summary.fitted.values[indd3r, c(1:3, 5)]
##                             mean     sd 0.025quant 0.975quant
## fitted.APredictor.201    9.785 0.3409      9.118     10.456
## fitted.APredictor.202   12.574 0.6362     11.327     13.827
## fitted.APredictor.203    6.750 1.0125      4.769      8.752
```

Also, the posterior marginal distribution of the predicted values can be obtained:

```
marg3r <- r5pmu$marginals.fitted.values[indd3r]
```

Finally, a 95% HPD interval for μ at the second target location can be computed with the function `inla.hpdmarginal()`:

```
inla.hpdmarginal(0.95, marg3r[[2]])
##                  low  high
## level:0.95 11.32 13.82
```

It is possible to see that around point (0.5, 0.5) the values of the response are significantly larger than β_0, as seen in Figure 2.10.

2.5.2 Summing the linear predictor components

A computationally cheap approach is to (naïvely) sum the projected posterior means of the terms in the linear regression. For covariates, this is done by considering the posterior means of the covariate coefficients and multiplying them by a covariate scenario, as done in Cameletti et al. (2013). In this toy example, we have only to consider the posterior mean of the intercept and the posterior mean of the random field:

```
res5$summary.fix[1, "mean"] +
  drop(A5pts3 %*% res5$summary.random$i$mean)
## [1]  9.785 12.574  6.751
```

For the standard error, a similar approach is possible. However, it needs to

account for the intercept variance and covariance of the two terms in the sum
as well.

2.5.3   Prediction on a grid

The computation of all marginal posterior distributions at the points of a grid
is computationally expensive. However, complete marginal distributions are
seldom used although they are computed by default. Instead, posterior means
and standard deviations are enough in most of the cases. As we do not need
the entire posterior marginal distribution at each point in the grid, we set an
option in the call to stop returning it in the output. This is useful for saving
memory when we want to store the object for use in the future.

In the code below, the model is fitted again, considering the mode for all
the hyperparameters in the model fitted previously, but the storage of the
marginal posterior distributions of random effects and posterior predictor
values is disabled (using parameter control.results). The computation of
the quantiles is also disabled, by setting quantiles equal to FALSE. Only the
mean and standard deviation are stored. Furthermore, the projector matrix
is the same that was used in the previous example to project the posterior
mean on the grid. Thus, we will have the mean and standard deviation of the
posterior marginal distribution of μ at each point in the grid.

```
stkgrid <- inla.stack(
  data = list(resp = NA),
  A = list(pgrid0$proj$A, 1),
  effects = list(i = 1:spde5$n.spde, beta0 = rep(1, 101 * 101)),
  tag = 'prd.gr')

stk.all <- inla.stack(stk5, stkgrid)

res5g <- inla(resp ~ 0 + beta0 + f(i, model = spde5),
  data = inla.stack.data(stk.all),
  control.predictor = list(A = inla.stack.A(stk.all),
    compute = TRUE),
  control.mode=list(theta=res5$mode$theta, restart = FALSE),
  quantiles = NULL,
  control.results = list(return.marginals.random = FALSE,
    return.marginals.predictor = FALSE))
```

Again, the index of the predicted values needs to be obtained:

```
igr <- inla.stack.index(stk.all, 'prd.gr')$data
```

This index can be used to visualize or summarize the predicted values. These predicted values have been plotted together with the prediction of the random field obtained in Section 2.4.1 in Figure 2.10.

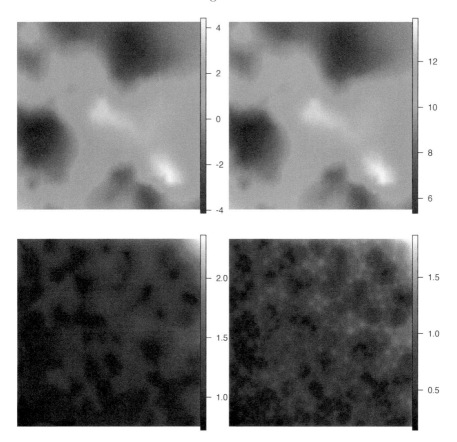

FIGURE 2.10 The mean and standard deviation of the random field (top left and bottom left, respectively) and the mean and standard deviation of the fitted values (top right and bottom right, respectively).

Figure 2.10 shows that there is a variation from -4 to 4 in the spatial effect. Considering also that standard deviations range from about 0.8 to 1.6, spatial dependence is considerable. In addition, the standard deviation of both the random field and μ are smaller near corner $(0, 0)$ and larger near corner $(1, 1)$. This is just proportional to the locations density.

The two standard deviation fields are different because one is for the random field only and the other is for the expected value of the outcome, which takes into account the standard deviation of the mean as well.

2.5.4   Results from different meshes

In this section, results for the toy dataset based on the six different meshes
built in Section 2.6 will be compared. To do this comparison, the posterior
marginal distributions of the model parameters will be plotted. The true values
used in the simulation of the toy dataset have been added to the plots in order
to evaluate the impact of the meshes on the results. Also, maximum likelihood
estimates using the geoR package (Ribeiro Jr. and Diggle, 2001) have been
added.

Six models will be fitted, using each one of the six meshes, and the results are
put together in a list using the following code:

```
lrf <- list()
lres <- list()
l.dat <- list()
l.spde <- list()
l.a <- list()

for (k in 1:6) {
  # Create A matrix
  l.a[[k]] <- inla.spde.make.A(get(paste0('mesh', k)),
    loc = coords)
  # Creeate SPDE spatial effect
  l.spde[[k]] <- inla.spde2.pcmatern(get(paste0('mesh', k)),
    alpha = 2, prior.range = c(0.1, 0.01),
    prior.sigma = c(10, 0.01))

  # Create list with data
  l.dat[[k]] <- list(y = SPDEtoy[,3], i = 1:ncol(l.a[[k]]),
    m = rep(1, ncol(l.a[[k]])))
  # Fit model
  lres[[k]] <- inla(y ~ 0 + m + f(i, model = l.spde[[k]]),
    data = l.dat[[k]], control.predictor = list(A = l.a[[k]]))
}
```

Mesh size influences the computational time needed to fit the model. More
nodes in the mesh need more computational time. The times required to fit
the models with INLA for these six meshes are shown in Table 2.1.

TABLE 2.1: Times (in seconds) required to build a mesh.

Mesh	Size	Time
1	2903	16.236
2	489	2.114

Mesh	Size	Time
3	371	1.387
4	2925	16.729
5	489	2.638
6	374	1.853

Furthermore, the posterior marginal distribution of σ_e^2 is computed for each fitted model:

```
s2.marg <- lapply(lres, function(m) {
  inla.tmarginal(function(x) exp(-x),
    m$internal.marginals.hyperpar[[1]])
})
```

The true values of the model parameters are: $\beta_0 = 10$, $\sigma_e^2 = 0.3$, $\sigma_x^2 = 5$, range $\sqrt{8}/7$ and $\nu = 1$. The ν parameter is fixed at the true value as a result of setting $\alpha = 2$ in the definition of the SPDE model.

```
beta0 <- 10
sigma2e <- 0.3
sigma2u <- 5
range <- sqrt(8) / 7
nu <- 1
```

The maximum likelihood estimates obtained with package geoR (Ribeiro Jr. and Diggle, 2001) are:

```
lk.est <- c(beta = 9.54, s2e = 0.374, s2u = 3.32,
  range = 0.336 * sqrt(8))
```

The posterior marginal distributions of β_0, σ_e^2, σ_x^2, and range have been plotted in Figure 2.11. Here, it can be seen how the posterior marginal distribution of the intercept has its mode at the likelihood estimate, considering the results from all six meshes.

The main differences can be found in the noise variance σ_e^2 (i.e., the nugget effect). The results from the mesh based on the points have a posterior mode which is smaller than the actual value and the maximum likelihood estimate. In general, as the number of triangles in the mesh increase, the posterior mode gets closer to the actual value of the parameter. Regarding the meshes computed from the boundary, they seem to provide posterior modes which are closer to the maximum likelihood estimate than the other meshes.

Regarding the marginal variance of the latent field, σ_x^2, all meshes produced results in a posterior smaller than the actual value and closer to the maximum likelihood estimate. For the range, all meshes have a mode smaller than the maximum likelihood estimate and very close to the actual value.

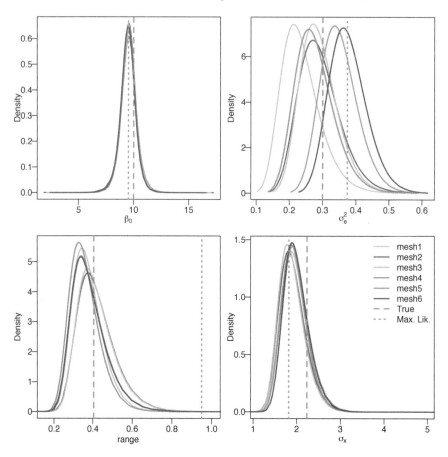

FIGURE 2.11 Marginal posterior distribution for β_0 (top left), σ_e^2 (top right), range (bottom left) and $\sqrt{\sigma_x^2}$ (bottom right).

Although these results are based on a toy example and are not conclusive, a general comment is that it is good to have a mesh tuned about the point locations, to capture noise variance. It is also important to allow for some flexibility to avoid a large variability in the shape and size of the triangles, to get good estimates of the latent field parameters. Thus, it is a tradeoff between (1) point locations and (2) large variability in the shape of the mesh triangles.

2.6 Triangulation details and examples

As stated earlier in this chapter, the first step to fit an SPDE model is the construction of the *mesh* to represent the spatial process. This step must be done very carefully, since it is similar to choosing the integration points on a numeric integration algorithm. Should the points be taken at regular intervals? How many points are needed? Furthermore, additional points around the boundary, or outer extension, need to be chosen with care. This is necessary to avoid a boundary effect that may cause a variance twice as larger at the border than within the domain (Lindgren, 2012; Lindgren and Rue, 2015).

Finally, in order to experiment with mesh building, there is a Shiny application in the INLA package that can be loaded with `meshbuilder()`. This will be useful to learn and understand how the different parameters are used in the definition of the mesh work. This is further described in Section 2.7.

2.6.1 Getting started

For a two dimensional mesh, function `inla.mesh.2d()` is the one that is recommended to use for building a mesh. This function creates a Constrained Refined Delaunay Triangulation (CRDT) over the study region, that will be simply referred to as the *mesh*. This function can take several arguments:

```
str(args(inla.mesh.2d))
## function (loc = NULL, loc.domain = NULL, offset = NULL,
##     n = NULL, boundary = NULL, interior = NULL,
##     max.edge = NULL, min.angle = NULL, cutoff = 1e-12,
##     max.n.strict = NULL, max.n = NULL, plot.delay = NULL,
##     crs = NULL)
```

First of all, some information about the study region is needed to create the mesh. This can be provided by the location points or just a domain. The point locations, which can be passed to the function in the `loc` argument, are used as initial triangulation nodes. A single polygon can be supplied to determine the domain extent using the `loc.domain` argument. After the location points or the boundary have been passed to the function, the algorithm will find a convex hull mesh. A non-convex hull mesh can be made when a list of polygons is passed using the `boundary` argument, where each element of this list is of the class returned by function `inla.mesh.segment()`. Hence, one of these three arguments (`loc`, `domain` or `boundary`) is mandatory.

The other mandatory argument is `max.edge`. This argument specifies the maximum allowed triangle edge length in the inner domain and in the outer extension. So, it can be a single numeric value or length two vector, and it must be in the same scale units as the coordinates.

The other arguments are used to specify additional constraints on the mesh. The offset argument is a numeric value (or a length two vector) and it is used to set the automatic extension distance. If positive, it is the extension distance in the same scale units. If negative, it is interpreted as a factor relative to the approximate data diameter; i.e., a value of -0.10 (the default) will add a 10% of the data diameter as outer extension.

Argument n is the initial number of points in the extended boundary. The interior argument is a list of segments to specify interior constraints, each one of the inla.mesh.segment class. A good mesh needs to have triangles as regular as possible in size and shape. This can be controlled with argument max.edge to control the edge length of the triangles and the min.angle argument (which can be scalar or length two vector) can be used to specify the minimum internal angles of the triangles in the inner domain and the outer extension. Values up to 21 guarantee the convergence of the algorithm (de Berg et al., 2008; Guibas et al., 1992).

To further control the shape of the triangles, the cutoff argument can be used to set the minimum allowed distance between points. It means that points at a closer distance than the supplied value are replaced by a single vertex. Hence, it avoids small triangles and must be a positive number, and is critical when there are some points very close to each other, either for point locations or in the domain boundary.

To understand how function inla.mesh.2d() works, several meshes have been computed for different combinations of some of the arguments using the first five locations of the toy dataset. First, the toy dataset will be loaded and its first five points taken:

```
data(SPDEtoy)
coords <- as.matrix(SPDEtoy[, 1:2])
p5 <- coords[1:5, ]
```

As the meshes will be built using the domain and not the points, this domain needs to be defined first:

```
pl.dom <- cbind(c(0, 1, 1, 0.7, 0, 0), c(0, 0, 0.7, 1, 1, 0))
```

This domain has been plotted in green over some of the meshes in Figure 2.12.

Finally, the meshes are created using the first five points or the domain defined above:

```
m1 <- inla.mesh.2d(p5, max.edge = c(0.5, 0.5))
m2 <- inla.mesh.2d(p5, max.edge = c(0.5, 0.5), cutoff = 0.1)
m3 <- inla.mesh.2d(p5, max.edge = c(0.1, 0.5), cutoff = 0.1)
m4 <- inla.mesh.2d(p5, max.edge = c(0.1, 0.5),
   offset = c(0, -0.65))
m5 <- inla.mesh.2d(loc.domain = pl.dom, max.edge = c(0.3, 0.5),
   offset = c(0.03, 0.5))
m6 <- inla.mesh.2d(loc.domain = pl.dom, max.edge = c(0.3, 0.5),
   offset = c(0.03, 0.5), cutoff = 0.1)
m7 <- inla.mesh.2d(loc.domain = pl.dom, max.edge = c(0.3, 0.5),
   n = 5, offset = c(0.05, 0.1))
m8 <- inla.mesh.2d(loc.domain = pl.dom, max.edge = c(0.3, 0.5),
   n = 7, offset = c(0.01, 0.3))
m9 <- inla.mesh.2d(loc.domain = pl.dom, max.edge = c(0.3, 0.5),
   n = 4, offset = c(0.05, 0.3))
```

These nine meshes have been displayed in Figure 2.12. In the next paragraphs we provide a comparative critical assessment of the quality of the different meshes depending on their structure.

The m1 mesh has two main problems. First, there are some triangles with small inner angles and, second, some large triangles appear in the inner domain. In the m2 mesh, the restriction on the locations is relaxed, because points with distance less than the cutoff are considered a single vertex. This avoids some of the triangles (at the bottom right corner) with small angles that appeared in mesh m1. Hence, using a cutoff is a very good idea. Each inner triangle in the m3 mesh has an edge length less than 0.1 and this mesh looks better than the two previous ones.

The m4 mesh has been made without first building a convex hull extension around the points. It has just the second outer boundary. In this case, the length of the inner triangles is not good (first value in the max.edge argument) and there are triangles with edge lengths of up to 0.5 in the outer region. The shape of the triangles looks good in general.

The m5 mesh was made just using the domain polygon and it has a shape similar to the domain area. In this mesh there are some small triangles at the corners of the domain due to the fact that it has been built without specifying a cutoff. Also, there is a (relatively) small first extension and a (relatively) large second one. In the m6 mesh, a cutoff has been added and a better mesh than the previous one has been obtained.

In the last three meshes, the initial number of extension points has been changed. It can be useful to change this value in some situations to get a convergence of the mesh building algorithm. For example, Figure 2.12 shows the shape of the mesh obtained with n = 5 in the m7 mesh. This number

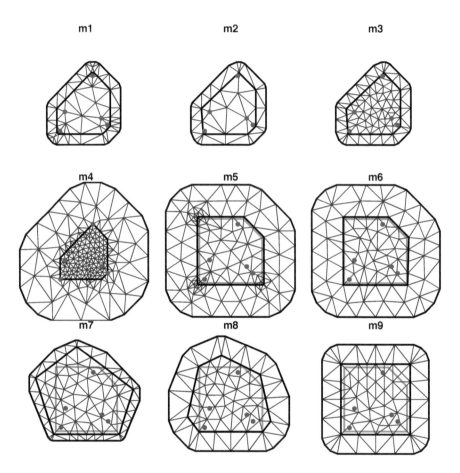

FIGURE 2.12 Different triangulations created with function `inla.mesh.2d()` using different combinations of its arguments.

produces a mesh that seems inadequate for this domain because there is a non uniform extension behind the border. Meshes m8 and m9 have very bad triangle shapes.

The object returned by the `inla.mesh.2d()` function is of class `inla.mesh` and it is a list with several elements:

```
class(m1)
## [1] "inla.mesh"
names(m1)
## [1] "meta"     "manifold" "n"        "loc"      "graph"
## [6] "segm"     "idx"      "crs"
```

The number of vertices in each mesh is in the n element of this list, which allows us to compare the number of vertices of all the meshes created:

```
c(m1$n, m2$n, m3$n, m4$n, m5$n, m6$n, m7$n, m8$n, m9$n)
## [1]   61   36   78 196 121   88   87  68   72
```

The **graph** element represents the CRDT obtained. In addition, the **graph** element contains the matrix that represents the graph of the neighborhood structure. For example, for mesh m1 this matrix has the following dimension:

```
dim(m1$graph$vv)
## [1] 61 61
```

The vertices that correspond to the location points are identified in the `idx` element:

```
m1$idx$loc
## [1] 24 25 26 27 28
```

2.6.2 Non-convex hull meshes

All meshes in Figure 2.12 have been made to have a convex hull boundary. In this context, a convex hull is a polygon of triangles out of the domain area, the extension made to avoid the boundary effect. A triangulation without an additional border can be made by supplying the **boundary** argument instead of the **location** or **loc.domain** arguments. One way to create a non-convex hull is to build a boundary for the points and supply it in the **boundary** argument.

Boundaries can also be created by using the `inla.nonconvex.hull()` function, which takes the following arguments:

```
str(args(inla.nonconvex.hull))
## function (points, convex = -0.15, concave = convex,
##      resolution = 40, eps = NULL, crs = NULL)
```

When using this function, the points and a set of constraints need to be passed. The shape of the boundary can be controlled, including its convexity, concavity and resolution. In the next example, some boundaries are created first and then a mesh is built with each one to better understand how this process works:

```
# Boundaries
bound1 <- inla.nonconvex.hull(p5)
bound2 <- inla.nonconvex.hull(p5, convex = 0.5, concave = -0.15)
bound3 <- inla.nonconvex.hull(p5, concave = 0.5)
bound4 <- inla.nonconvex.hull(p5, concave = 0.5,
  resolution = c(20, 20))

# Meshes
m10 <- inla.mesh.2d(boundary = bound1, cutoff = 0.05,
  max.edge = c(0.1, 0.2))
m11 <- inla.mesh.2d(boundary = bound2, cutoff = 0.05,
  max.edge = c(0.1, 0.2))
m12 <- inla.mesh.2d(boundary = bound3, cutoff = 0.05,
  max.edge = c(0.1, 0.2))
m13 <- inla.mesh.2d(boundary = bound4, cutoff = 0.05,
  max.edge = c(0.1, 0.2))
```

These meshes have been displayed in Figure 2.13 and a discussion on the quality of the produced meshes is provided below.

The m10 mesh is built with a boundary that uses the default arguments in the inla.nonconvex.hull() function. The default convex and concave arguments are both equal to a proportion of 0.15 of the points' domain radius, that is computed as follows:

```
0.15 * max(diff(range(p5[, 1])), diff(range(p5[, 2])))
## [1] 0.1291
```

If we supply a larger value in the convex argument, such as the one used to generate mesh m11, a larger boundary is obtained. This happens because all circles with a center at each point and a radius less than the convex value are inside the boundary. When a larger value for the concave argument is taken, as in the boundary used for the m12 and m13 meshes, there are no circles with radius less than the concave value outside the boundary. If a smaller resolution

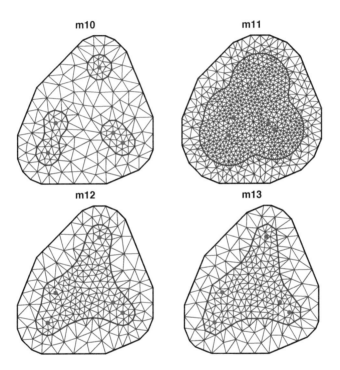

FIGURE 2.13 Non-convex meshes with different boundaries.

is chosen, a boundary with small resolution (in terms of number of points) is obtained. For example, compare the `m12` and `m13` meshes.

2.6.3   Meshes for the toy example

To analyze the toy data set, six triangulation sets of options were used to make comparisons in Section 2.5.4. The first and second meshes force the location points to be vertices of the mesh, but the maximum length of the edges is allowed to be larger in the second mesh:

```
mesh1 <- inla.mesh.2d(coords, max.edge = c(0.035, 0.1))
mesh2 <- inla.mesh.2d(coords, max.edge = c(0.15, 0.2))
```

The third mesh is based on the points, but a cutoff greater than zero is taken to avoid small triangles in regions with a high number of observations:

```
mesh3 <- inla.mesh.2d(coords, max.edge = c(0.15, 0.2),
   cutoff = 0.02)
```

Three other meshes have been created based on the domain area. These are built to have approximately the same number of vertices as the previous ones:

```
mesh4 <- inla.mesh.2d(loc.domain = pl.dom,
   max.edge = c(0.0355, 0.1))
mesh5 <- inla.mesh.2d(loc.domain = pl.dom,
   max.edge = c(0.092, 0.2))
mesh6 <- inla.mesh.2d(loc.domain = pl.dom,
   max.edge = c(0.11, 0.2))
```

These six meshes are shown in Figure 2.14. The number of nodes in each one of these meshes is:

```
c(mesh1$n, mesh2$n, mesh3$n, mesh4$n, mesh5$n, mesh6$n)
## [1] 2903  489  371 2925  489  374
```

2.6.4   Meshes for Paraná state

In this book there are some examples using data collected in Paraná state, Brazil. In order to build a mesh to represent this region, two things need to be taken into account: the shape of this domain area and its coordinate reference system.

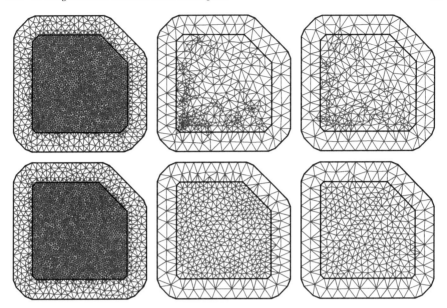

FIGURE 2.14 Six triangulations with different options for the toy example.

First of all, the daily rainfall dataset will be loaded (as this contains the boundary of Paraná state that can be used in the analysis):

```
data(PRprec)
```

This dataset comprises daily rainfall data from 616 stations in the 2011 year. The coordinates (two first columns) are in longitude and latitude. Altitude of the station is under the third column. Hence, the dimension of the data is:

```
dim(PRprec)
## [1] 616 368
```

The coordinates and data for the first 5 days of the year of the first two stations can be seen below to get an idea of the information available in this dataset:

```
PRprec[1:2, 1:8]
##    Longitude  Latitude  Altitude  d0101  d0102  d0103  d0104  d0105
## 1     -50.87    -22.85       365      0      0      0      0      0
## 3     -50.77    -22.96       344      0      1      0      0      0
```

Also, boundaries of Paraná state as a polygon are available:

```
data(PRborder)
dim(PRborder)
## [1] 2055    2
```

This is a set of 2055 points, in longitude and latitude.

As the boundary is irregular, in this case it is best to use a non-convex hull mesh. The first step is to build a non-convex domain using the locations of the precipitation data, as follows:

```
prdomain <- inla.nonconvex.hull(as.matrix(PRprec[, 1:2]),
  convex = -0.03, concave = -0.05,
  resolution = c(100, 100))
```

Using this defined domain, two meshes will be built with different resolutions (i.e., different maximum edge lengths) in the inner domain:

```
prmesh1 <- inla.mesh.2d(boundary = prdomain,
  max.edge = c(0.7, 0.7), cutoff = 0.35,
    offset = c(-0.05, -0.05))
prmesh2 <- inla.mesh.2d(boundary = prdomain,
  max.edge = c(0.45, 1), cutoff = 0.2)
```

The number of points in each mesh is 187 and 381, respectively. Both meshes have been displayed in Figure 2.15.

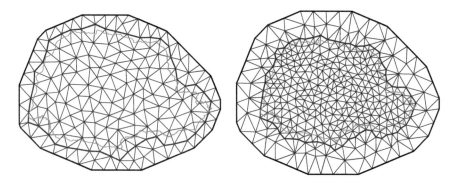

FIGURE 2.15 Meshes for Paraná state (Brazil).

2.6.5   Triangulation with a `SpatialPolygonsDataFrame`

Sometimes, the domain region is available as a map in some GIS format. In R, a widely used representation of a spatial object is made using object classes in the sp package (Bivand et al., 2013). To show an application in this case, the North Carolina map, which is extensively analyzed in the examples in package spdep (Bivand and Piras, 2015), will be used. Shapefiles of North Carolina are currently available in package spData (Bivand et al., 2018).

First of all, the map will be loaded using function readOGR() from package rgdal (Bivand et al., 2017):

```
library(rgdal)
# Get filename of file to load
nc.fl <- system.file("shapes/sids.shp", package = "spData")[1]
# Load shapefile
nc.sids <- readOGR(strsplit(nc.fl, 'sids')[[1]][1], 'sids')
```

This map contains the boundaries of the different counties in North Carolina. For this reason, this map is simplified by uniting all the county areas together in order to get the outer boundary of North Carolina. To do it, the gUnaryUnion() function from the rgeos package (Bivand and Rundel, 2017) is used:

```
library(rgeos)
nc.border <- gUnaryUnion(nc.sids, rep(1, nrow(nc.sids)))
```

Now, the inla.sp2segment() function is used to extract the boundary of the SpatialPolygons object that contains the boundary of the map:

```
nc.bdry <- inla.sp2segment(nc.border)
```

Then, the mesh is finally created and it has been displayed in Figure 2.16:

```
nc.mesh <- inla.mesh.2d(boundary = nc.bdry, cutoff = 0.15,
   max.edge = c(0.3, 1))
```

2.6.6 Mesh with holes and physical boundaries

So far, the domains considered in the analysis were simply connected because they did not contain any holes. However, sometimes we may find that the spatial process under study occurs in a domain which contain holes. These holes also produce physical boundaries that need to be taken into account in the analysis.

An example of application is when modeling fish and inland is required to be considered as a physical barrier. In addition, islands in the study region will be required to be taken into account. Similarly, when studying species

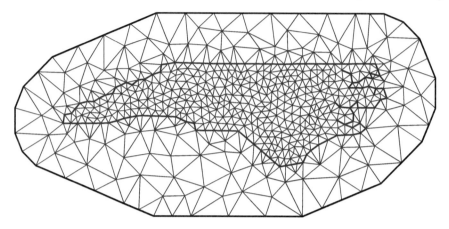

FIGURE 2.16 Mesh constructed using the North Carolina map.

distribution in a forest, lakes and rivers may introduce natural barriers that also need to be taken into account. These types of models are further described in Chapter 5.

The polygons in Figure 2.17 illustrate this case, as they include a hole in the left polygon. These polygons have been created with the following commands:

```
pl1 <- Polygon(cbind(c(0, 15, 15, 0, 0), c(5, 0, 20, 20, 5)),
   hole = FALSE)
h1 <- Polygon(cbind(c(5, 7, 7, 5, 5), c(7, 7, 15, 15, 7)),
   hole = TRUE)
pl2 <- Polygon(cbind(c(15, 20, 20, 30, 30, 15, 15),
   c(10, 10, 0, 0, 20, 20, 10)), hole = FALSE)
sp <- SpatialPolygons(
   list(Polygons(list(pl1, h1), '0'), Polygons(list(pl2), '1')))
```

Here, `pl1` corresponds to the left polygons, with a hole defined by polygon `h1`, and `pl2` is the polygon on the right. All these three polygons are combined into a `SpatialPolygons` object called `sp`.

Hence, there are two neighboring regions, one with a hole and another one with no convex shape. Let us consider that the two polygons represent two regions in a lake with their outer boundaries representing the shore and with the hole on the left representing an island. Suppose that it is necessary to avoid correlation between near regions separated by land. For example, suppose that correlation between A and C should be smaller than between A and B or between B and C.

In this example, additional points for the outer boundary are not needed. For

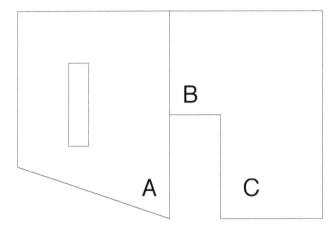

FIGURE 2.17 Region with a hole and non-convex domain.

this reason, a length one value for `max.edge` must be passed. The following code prepares the boundary and builds the mesh:

```
bound <- inla.sp2segment(sp)
mesh <- inla.mesh.2d(boundary = bound, max.edge = 2)
```

The mesh is displayed in Figure 2.18. Notice that when building the SPDE model, the neighborhood structure of the mesh is taken into account. In this case, it is easier to reach B from A than to reach B from C on the related graph. Hence, the SPDE model will consider a higher correlation between B and A than between B and C.

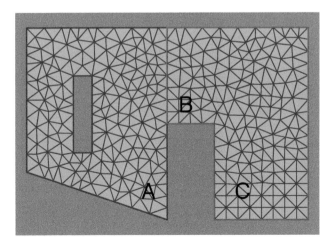

FIGURE 2.18 Triangulation with hole and a non-convex region.

2.7   Tools for mesh assessment

Building the right mesh for a spatial model may require some tuning and time. For this reason, INLA includes a Shiny (Chang et al., 2018) application that can be used to set the parameters interactively and display the resulting mesh in a few seconds. Figure 2.19 shows the Shiny application, which can be run with:

```
meshbuilder()
```

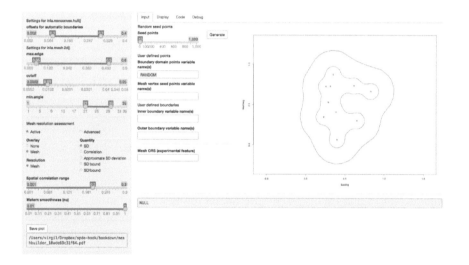

FIGURE 2.19 Shiny application to interactively create a mesh obtained with command `meshbuilder()`.

The mesh builder has a number of options to define the mesh on the left side. These include options to be passed to functions `inla.nonconvex.hull()` and `inla.mesh.2d()` and the resulting mesh displayed on the right part. Below these options, there are a number of options to assess the mesh resolution. These include a number of quantities to be plotted, such as the standard deviation (SD).

The right part of the application has four tabs that can be clicked to show different outputs. By default, the application opens the `Input` tab, which includes options to set the points and, possibly, boundaries to create the mesh.

The `Display` tab shows the created mesh, as well as some information about it at the bottom. If option `Active` under `Mesh resolution assessment` has been clicked, then this plot will also show the variable marked under `Quantity`.

By selecting different quantities, it is possible to assess how the created mesh will produce the estimates from the model. Figure 2.20 shows the generated mesh as well as the point values of the model standard deviation. Note how the higher values appear in the outer region, which is close to the mesh boundary and outside the domain of interest. By including this outer region the estimates of the spatial process in the study region do not suffer from any boundary effect.

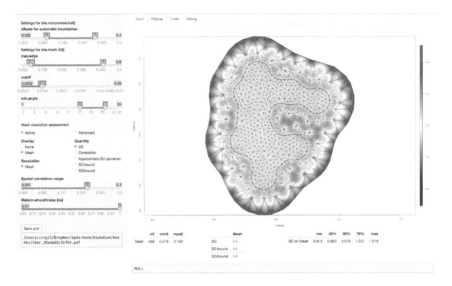

FIGURE 2.20 Display of the estimated standard deviation of the spatial process under the `Display` tab in the Shiny application.

The next tab, is the `Code` tab. This will show the `R` code used to generate the mesh. Finally, the `Debug` tab provides some internal debugging information.

2.8   Non-Gaussian response: Precipitation in Paraná

Now, more complex spatial models based on the SPDE approach will be considered. In particular, the analysis of non-Gaussian observations, which will be illustrated using rainfall data from Paraná (Brazil).

Climate data is a very common type of spatial data. The example data we will use here is collected by the National Water Agency in Brazil (*Agencia Nacional de Águas*, ANA, in Portuguese). ANA collects data from many locations over Brazil, and all these data are freely available from the ANA website (`http://www3.ana.gov.br/portal/ANA`).

2.8.1   The data set

As mentioned in Section 2.6.4, this dataset is composed of daily rainfall data
from year 2011 at 616 locations, including stations within Paraná state and
around its border. This dataset is available in the INLA package and it can be
loaded as follows:

```
data(PRprec)
```

The coordinates of the locations where the data were collected are in the first
two columns, altitude is next in the third column and the following 365 columns,
one for each day of the year, store the daily accumulated precipitation.

In the code below the first eight columns from four stations of the PRprec
dataset are shown. These four stations (defined in index ii) are the one with
the lowest latitude (among all stations with missing altitude), the ones with
the lowest and largest longitude and the one with the highest altitude.

```
ii <- c(537, 304, 610, 388)
PRprec[ii, 1:8]
##        Longitude Latitude Altitude d0101 d0102 d0103 d0104 d0105
## 1239    -48.94   -26.18      NA    20.5   7.9   8.0   0.8   0.1
## 658     -48.22   -25.08       9    18.1   8.1   2.3  11.3  23.6
## 1325    -54.48   -25.60     231    43.8   0.2   4.6   0.4   0.0
## 885     -51.52   -25.73    1446     0.0  14.7   0.0   0.0  28.1
```

These four stations have been represented using red crosses in the right plot in
Figure 2.21.

There are a few problems with this dataset. First of all, there are seven stations
with missing altitude and missing data on daily rainfall, which are displayed
in red in the left plot in Figure 2.21. If altitude is considered when building a
model it will be important to have this variable measured everywhere in the
state. There are digital elevation models that can be considered to find out the
altitude at these locations. Alternatively, a stochastic model for this variable
can be considered.

Daily rainfall mean in January 2011 will be considered in the analysis. However,
there are 269 missing observations. As a simple imputation method, the average
over the number of days in January 2011 without missing data will be taken,
as follows:

```
PRprec$precMean <- rowMeans(PRprec[, 3 + 1:31], na.rm = TRUE)
summary(PRprec$precMean)
```

```
##     Min. 1st Qu.  Median   Mean 3rd Qu.   Max.   NA's
##     0.50    4.97    6.54    6.96    8.63   21.66      6
```

There are still some stations with missing values of the monthly average in January 2011. The number of missing observations in July 2011 at each station can be checked as:

```
table(rowSums(is.na(PRprec[, 3 + 1:31])))
##
##    0    2    4    8   17   26   31
## 604    1    1    1    1    2    6
```

Hence, there are some stations with all the observations missing during January 2011, and this is the reason why the average during January 2011 is also missing.

In addition to the rainfall data, the Paraná state border will also be considered to define the study domain:

```
data(PRborder)
```

The locations of the data are displayed in Figure 2.21. The size of the points on the left plot is proportional to the altitude at the locations. The cyan line in the east border is the coast (along the Atlantic Ocean). There are low altitudes near the ocean, and high altitudes around 50 to 100 kilometers from this coast and from the center of the state towards the south as well. Altitude decreases to the north and west sides of Paraná state. The size of the points in the right plot is proportional to the daily average of the precipitation in January 2011. In this case, there are higher values near the coast.

2.8.2 Model and covariate selection

In this section the average of the daily accumulated precipitation for each of the 31 days in January 2011 will be analyzed. Given that it must be a positive value, a Gamma likelihood will be considered. In the Gamma likelihood, the mean is $E(y_i) = a_i/b_i = \mu_i$ and the variance is $V(y_i) = a_i/b_i^2 = \mu_i^2/\phi$, where ϕ is a precision parameter. Then it is necessary to define a model for the linear predictor $\eta_i = \log(\mu_i)$, which will depend on the covariates $\mathbf{F}$ and the spatial random field $\mathbf{x}$ as follows:

$$
\begin{aligned}
y_i|\mathbf{F_i}, \alpha, x_i, \theta &\sim Gamma(a_i, b_i) \\
\log(\mu_i) &= \alpha + f(\mathbf{F_i}) + x_i \\
\mathbf{x} &\sim GF(0, \Sigma)
\end{aligned}
$$

Here, $\mathbf{F_i}$ is a vector of covariates (the location coordinates and altitude of

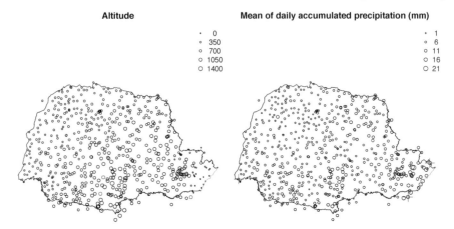

FIGURE 2.21 Locations of Paraná stations, altitude and average of daily accumulated precipitation (mm) in January 2011. Circles in red denote stations with missing observations and red crossess denote the four stations shown in the example in the main text.

station i) that will be assumed to follow a function detailed later and $\mathbf{x}$ is the spatial latent Gaussian random field (GF). For this, a Matérn covariance function will be considered with parameters ν, κ and σ_x^2.

Smoothed covariate effect

An initial exploration of the relationship between precipitation and the other covariates has been made by means of dispersion diagrams, available in Figure 2.22. After preliminary tests, it seems reasonable to construct a new covariate: the distance from each station to the Atlantic Ocean. The Paraná state border along the Atlantic Ocean is shown as a cyan line in Figure 2.21. The minimum distance from each station to this line can be computed to obtain the distance from the station to the Atlantic Ocean.

To have this distance in kilometers the `spDists()` function from the `sp` package will be used (with argument `longlat` equal to `TRUE`):

```
coords <- as.matrix(PRprec[, 1:2])
mat.dists <- spDists(coords, PRborder[1034:1078, ],
  longlat = TRUE)
```

However, this function computes the distance between each location in the first set of points to each point in the second set of points. So, the minimum value along the rows of the resulting matrix of distances must be taken to obtain the minimum distance to the ocean:

```
PRprec$oceanDist <- apply(mat.dists, 1, min)
```

The dispersion plots can be seen in Figure 2.22. The average precipitation seems to have a well defined non-linear relationship with longitude. Also, there is a similar, but inverse, relation with distance to the ocean. Hence, two models will be built. The first one will include longitude as a covariate and the second one will include distance to the ocean. Model fitting measures will be computed to proceed with model choice between these two options.

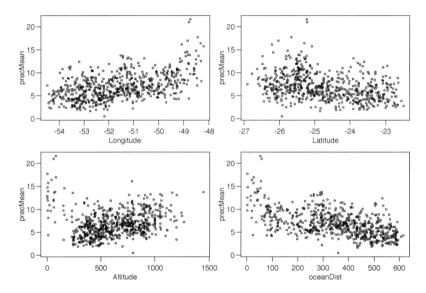

FIGURE 2.22 Dispersion plots of average of daily accumulated precipitation by longitude (top left), latitude (top right), altitude (bottom left) and distance to ocean (bottom right).

To consider a non-linear relationship for a covariate a random walk prior on its effect can be set. To do that, the covariate can be discretized into a set of knots and the random walk prior placed on them. In this case, the term in the linear predictor due to ocean distance (or longitude) is discretized into m classes considering the `inla.group()` function. The model can be chosen from any one dimensional model available in `INLA`, such as `rw1`, `rw2`, `ar1` or others. It can also be modeled using a one-dimensional Matérn model.

When considering intrinsic models as prior distributions, scaling the model should be considered (Sørbye and Rue, 2014). After scaling, the precision parameter can be interpreted as the inverse of the random effect marginal variance. Hence, scaling makes the process of defining a prior easier. The suggestion here is to consider the PC-prior (Simpson et al., 2017) introduced in Section 1.6.5. This can be done by defining a reference standard deviation

σ_0 and the right tail probability u as $P(\sigma > \sigma_0) = u$. In this example, these values will be set to $\sigma_0 = 1$ and $u = 0.01$. These values can be stored in a list (to be passed to `inla()` later) as follows:

```
pcprec <- list(prior = 'pcprec', param = c(1, 0.01))
```

Define the spatial model and prepare the data

In order to define the spatial model a mesh must be defined. First, a boundary around the points will be defined using a non-convex hull and used to create the mesh:

```
pts.bound <- inla.nonconvex.hull(coords, 0.3, 0.3)
mesh <- inla.mesh.2d(coords, boundary = pts.bound,
    max.edge = c(0.3, 1), offset = c(1e-5, 1.5), cutoff = 0.1)
```

Once the mesh has been defined, the projector matrix is computed:

```
A <- inla.spde.make.A(mesh, loc = coords)
```

The SPDE model considering the PC-prior derived in Fuglstad et al. (2018) for the model parameters as the practical range, $\sqrt{8\nu}/\kappa$, and the marginal standard deviation is defined as follows:

```
spde <- inla.spde2.pcmatern(mesh = mesh,
    prior.range = c(0.05, 0.01), # P(practic.range < 0.05) = 0.01
    prior.sigma = c(1, 0.01)) # P(sigma > 1) = 0.01
```

The stack data is defined to include four effects: the GF, the intercept, longitude and distance to the ocean. The R code to do this is:

```
stk.dat <- inla.stack(
  data = list(y = PRprec$precMean),
  A = list(A,1),
  effects = list(list(s = 1:spde$n.spde),
    data.frame(Intercept = 1,
      gWest = inla.group(coords[, 1]),
      gOceanDist = inla.group(PRprec$oceanDist),
      oceanDist = PRprec$oceanDist)),
  tag = 'dat')
```

Fitting the models

Both models will be fitted using the same data stack as only the formula needs to be changed. For the model with longitude, the R code for model fitting is:

```
f.west <- y ~ 0 + Intercept + # f is short for formula
  f(gWest, model = 'rw1', # first random walk prior
    scale.model = TRUE, # scaling this random effect
    hyper = list(theta = pcprec)) + # use the PC prior
  f(s, model = spde)

r.west <- inla(f.west, family = 'Gamma', # r is short for result
  control.compute = list(cpo = TRUE),
  data = inla.stack.data(stk.dat),
  control.predictor = list(A = inla.stack.A(stk.dat), link = 1))
```

Option `link = 1` in the `control.predictor` is used to set the function link to be considered in the computation of the fitted values that form the vector of available links for the Gamma likelihood:

```
inla.models()$likelihood$gamma$link
## [1] "default"  "log"       "quantile"
```

We recommend always using `link = 1` if you do not have several likelihoods. This will use the default link (i.e., the natural logarithm), but other link functions can be used by changing the value of `link`.

For the model with distance to the ocean, the necessary code is:

```
f.oceanD <- y ~ 0 + Intercept +
  f(gOceanDist, model = 'rw1', scale.model = TRUE,
    hyper = list(theta = pcprec)) +
  f(s, model = spde)

r.oceanD <- inla(f.oceanD, family = 'Gamma',
  control.compute = list(cpo = TRUE),
  data = inla.stack.data(stk.dat),
  control.predictor = list(A = inla.stack.A(stk.dat), link = 1))
```

In Figure 2.23 it can be seen how the effect from distance to the ocean is almost linear. Hence, another model considering this linear effect has been fitted:

```
f.oceanD.1 <- y ~ 0 + Intercept + oceanDist +
  f(s, model = spde)
r.oceanD.1 <- inla(f.oceanD.1, family = 'Gamma',
  control.compute = list(cpo = TRUE),
  data = inla.stack.data(stk.dat),
  control.predictor = list(A = inla.stack.A(stk.dat), link = 1))
```

Model comparison and results

The negative sum of the log CPO (Pettit, 1990, see Section 1.4) can be computed for each model using the following function:

```
slcpo <- function(m, na.rm = TRUE) {
  - sum(log(m$cpo$cpo), na.rm = na.rm)
}
```

Then, it can be used to compute this criterion for the three fitted models:

```
c(long = slcpo(r.west), oceanD = slcpo(r.oceanD),
  oceanD.1 = slcpo(r.oceanD.1))
##     long    oceanD oceanD.1
##     1279     1278     1274
```

These results suggest that the model with distance to the ocean as a linear effect has a better fit than the other two models. This is just one way to show how to compare models considering how they fit the data. In this case, the three models have a very similar sum of the log CPO.

For the model with the best fit, a summary of the posterior distribution of the intercept and the covariate coefficient can be obtained with:

```
round(r.oceanD.1$summary.fixed, 2)
##               mean   sd 0.025quant 0.5quant 0.975quant mode kld
## Intercept    2.43 0.09       2.25     2.43       2.62 2.42   0
## oceanDist    0.00 0.00       0.00     0.00       0.00 0.00   0
```

Similarly, summaries for the Gamma likelihood dispersion parameter can be obtained:

```
round(r.oceanD.1$summary.hyperpar[1, ], 3)
##                                                        mean     sd
## Precision parameter for the Gamma observations 14.88 1.45
```

```
##                                                      0.025quant
## Precision parameter for the Gamma observations           12.19
##                                                       0.5quant
## Precision parameter for the Gamma observations           14.83
##                                                     0.975quant
## Precision parameter for the Gamma observations           17.88
##                                                          mode
## Precision parameter for the Gamma observations 14.76
```

The summaries for the standard deviation and the practical range of the spatial process are:

```
round(r.oceanD.1$summary.hyperpar[-1, ], 3)
##                mean    sd 0.025quant 0.5quant 0.975quant   mode
## Range for s 0.669 0.143      0.440    0.650      1.001 0.613
## Stdev for s 0.234 0.022      0.193    0.233      0.280 0.231
```

Figure 2.23 displays the posterior marginals and estimates of some of the parameters in the previous models. For the model that considers a linear effect on the distance to the ocean, Figure 2.23 includes the posterior marginal distribution of the intercept β_0, the coefficient of the distance to the ocean, the precision of the Gamma likelihood, and the range (in degrees) and standard deviation of the spatial effect. In order to compare the smooth term on the distance to the ocean, the top-left plot in Figure 2.23 shows the estimate of this effect using a random walk of order 1. As can be seen, this effect looks like a linear effect, and this is the reason why the model that considers a linear effect on distance to the ocean provides a better fit.

If the significance of the spatial random effect component in the model needs to be assessed, a model without this term can be fitted. Then, the sum of the log CPO of these two models can be compared. A model without the spatial term can be fitted as follows:

```
r0.oceanD.1 <- inla(y ~ 0 + Intercept + oceanDist,
   family = 'Gamma', control.compute = list(cpo = TRUE),
   data = inla.stack.data(stk.dat),
   control.predictor = list(A = inla.stack.A(stk.dat), link = 1))
```

In order to compare the negative sum of the log CPO, the following R can be useful:

```
c(oceanD.1 = slcpo(r.oceanD.1), oceanD.10 = slcpo(r0.oceanD.1))
```

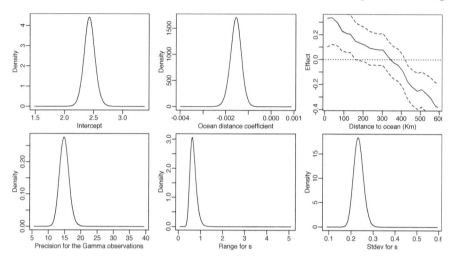

FIGURE 2.23 Posterior marginal distributions for β_0 (top left), the distance to ocean coefficient (top center), the Gamma likelihood precision (bottom left), the practical range (bottom center) and the standard deviation of the spatial field (bottom right). The top-right plot represents the posterior mean (continuous line) and 95% credibility interval (dashed lines) for the distance to ocean effect.

```
##   oceanD.l oceanD.l0
##      1274      1371
```

Given these results, it can be concluded that the best model is the one that includes the spatial term and a linear effect on the distance to the ocean.

2.8.3   Prediction of the random field

The spatial effect can be visualized by projecting it on a grid. A regular grid will be considered so that each pixel is a square with a side of about 4 kilometers. A step of 4/111 will be used because each degree has approximately 111 kilometers, so that the size of the pixels can be given in degrees. Then, the step size is defined as:

```
stepsize <- 4 * 1 / 111
```

As seen in Figure 2.21, the shape of Paraná state is wider along the x-axis than the y-axis and this fact will be considered when defining the grid. First, the range along each axis will be computed and then the size of the pixels will be used to set the number of pixels in each direction. We will divide the range along each axis by the step size and round it (to obtain integer values of the number of pixels in each direction):

```
x.range <- diff(range(PRborder[, 1]))
y.range <- diff(range(PRborder[, 2]))
nxy <- round(c(x.range, y.range) / stepsize)
```

Hence, the number of pixels in each direction is:

```
nxy
## [1] 183 117
```

The `inla.mesh.projector()` function will create a projector matrix. If a set of coordinates is not supplied it will create a grid automatically. In the next example, the limits and dimension of the grid are set as desired to match the boundaries of Paraná state:

```
projgrid <- inla.mesh.projector(mesh, xlim = range(PRborder[, 1]),
  ylim = range(PRborder[, 2]), dims = nxy)
```

Then, this can be passed on to the `inla.mesh.project()` function to do the projection of the posterior mean and the posterior standard deviation:

```
xmean <- inla.mesh.project(projgrid,
  r.oceanD$summary.random$s$mean)
xsd <- inla.mesh.project(projgrid, r.oceanD$summary.random$s$sd)
```

To improve the visualization, the pixels that fall outside Paraná state will be assigned a value of `NA`. Points outside the state boundaries can be found with function `inout()` from the `splancs` package (Rowlingson and Diggle, 1993, 2017), as follows:

```
library(splancs)
xy.in <- inout(projgrid$lattice$loc, PRborder)
```

Points with a value of `TRUE` will be those inside the boundaries of Paraná state. The number of points in the grid inside and outside can be computed with:

```
table(xy.in)
## xy.in
## FALSE  TRUE
##  7865 13546
```

Finally, the points that fall outside the boundaries are set to have a missing value of the posterior mean and standard deviation:

```
xmean[!xy.in] <- NA
xsd[!xy.in] <- NA
```

The posterior mean and posterior standard deviation for the spatial effect at each pixel can be seen in Figure 2.24. In the top left plot it can be seen that the posterior mean varies from -0.6 to 0.5. This is the variation after accounting for the effect of the distance to the ocean. When comparing it to the standard deviation plot, these values seem to be considerable as the standard deviations takes a maximum value of about 0.2. The variation in the standard deviation is mainly due to the density of the stations over the region. At the top right plot in Figure 2.24, the first zone of high values from right to left is near the capital city of Curitiba, where the number of stations around is relatively larger than in any other regions of the state.

2.8.4   Prediction of the response on a grid

When the objective is to predict the response, a simple approach will be to add all the other terms to the spatial field projected in the previous section and apply the inverse of the link function. A full Bayesian analysis for this problem will involve computing the joint prediction of the response together with the estimation of the model, as in the first example below. However, it can be computationally expensive when the number of pixels in the grid is large. For this reason, we consider different options here, including a cheap way for the specific case in the second example below.

By computation of the posterior distributions

Considering the grid over Paraná state from the previous section, it can be avoided to compute the posterior marginal distributions at those pixels that are not inside the state boundaries. That is, only the lines in the projector matrix corresponding to points inside the state will be considered:

```
Aprd <- projgrid$proj$A[which(xy.in), ]
```

The covariates for each pixel in the grid are needed now. In order to have them, the coordinates can be extracted from the projector object as well:

```
prdcoo <- projgrid$lattice$loc[which(xy.in), ]
```

Computing the distance to the ocean for each selected pixel is easy:

```
oceanDist0 <- apply(spDists(PRborder[1034:1078, ],
  prdcoo, longlat = TRUE), 2, min)
```

Suppose that the model to be fitted is the one with the smoothed effect of distance to the ocean. Each computed distance needs to be discretized in the same way as for the estimation data at the stations. First of all, the used knots need to be obtained and ordered:

```
OcDist.k <- sort(unique(stk.dat$effects$data$gOceanDist))
```

The knots are actually the middle values of the breaks and they will be considered as follows:

```
oceanDist.b <- (OcDist.k[-1] + OcDist.k[length(OcDist.k)]) / 2
```

Next, each distance to the ocean computed for a pixel needs to be matched to a knot and this knot assigned to the pixel:

```
i0 <- findInterval(oceanDist0, oceanDist.b) + 1
gOceanDist0 <- OcDist.k[i0]
```

Building the data stack with the prediction data and joining it to the data with the data for estimation can be done as follows:

```
stk.prd <- inla.stack(
  data = list(y = NA),
  A = list(Aprd, 1),
  effects = list(s = 1:spde$n.spde,
    data.frame(Intercept = 1, oceanDist = oceanDist0)),
  tag = 'prd')
stk.all <- inla.stack(stk.dat, stk.prd)
```

When fitting the model for prediction, the mode of the hyperparameters (i.e., `theta`) obtained in the previous model can be passed as known values to the new `inla()` call. It is also possible to avoid computing quantities not needed, such as quantiles, and avoid returning objects not needed, such as the posterior marginal distributions for the random effects and the linear predictor. Since the number of latent variables is the main issue in this case, the adaptive approximation will reduce computation time. It can be set using `control.inla = list(strategy = 'adaptive')`. Only random effects will get an adaptive approximation; any fixed effects are still approximated with the simplified Laplace method.

Hence, the code to fit the model with the previous settings is:

```
r2.oceanD.l <- inla(f.oceanD.l, family = 'Gamma',
  data = inla.stack.data(stk.all),
  control.predictor = list(A = inla.stack.A(stk.all),
    compute = TRUE, link = 1),
  quantiles = NULL,
  control.inla = list(strategy = 'adaptive'),
  control.results = list(return.marginals.random = FALSE,
    return.marginals.predictor = FALSE),
  control.mode = list(theta = r.oceanD.l$mode$theta,
    restart = FALSE))
```

As in other examples, it is necessary to find the indices for the predictions in the grid. It can be done by using the `inla.stack.index()` function. These values must be assigned into the right positions of a matrix with the same dimension as the grid. This is required for both the posterior predicted mean and its standard deviation:

```
id.prd <- inla.stack.index(stk.all, 'prd')$data
sd.prd <- m.prd <- matrix(NA, nxy[1], nxy[2])
m.prd[xy.in] <- r2.oceanD.l$summary.fitted.values$mean[id.prd]
sd.prd[xy.in] <- r2.oceanD.l$summary.fitted.values$sd[id.prd]
```

The results are displayed in the bottom row of Figure 2.24.

The posterior mean for the expected rainfall in January 2011 was higher near the ocean (to the east) and lower in the north-west side of Paraná state. Since the linear effect from the distance to the ocean will drive this pattern, the spatial effect in this case will account for deviations from this pattern. That is, the spatial effect is higher near the ocean and will add up to the higher effect from distance to the ocean, so that locations near the coast will have a higher posterior mean. The spatial effect is also higher in a region in the west, causing the expected values to be higher than what would be predicted by the linear effect alone.

Sampling at mesh nodes and interpolating

When all the covariates are smooth over space, it makes sense to make the predictions at the mesh nodes, where the spatial effect is estimated, and then project onto the grid. However, if a covariate is not smooth over space this approach no longer makes sense. The advantage of this approach is that it is computationally cheaper than to compute the full posterior marginals as in the previous examples.

In this example the idea is to compute the distance to the ocean effect at the

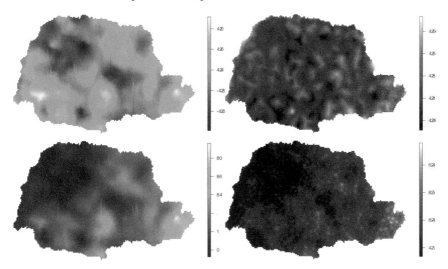

FIGURE 2.24 Posterior mean and standard deviation of the random field (top left and top right, respectively). Posterior mean and standard deviation for the response (bottom left and bottom right, respectively).

mesh nodes. Then, compute the linear predictor at the mesh nodes. By that, it is possible to predict the response at the mesh nodes and then to interpolate it.

The first step is to compute the enviromental covariate at the mesh nodes:

```
oceanDist.mesh <- apply(
  spDists(PRborder[1034:1078, ], mesh$loc[, 1:2], longlat = TRUE),
  2, min)
```

The method presented now is based on getting samples from the posterior distribution of the linear predictor at the mesh nodes. Then, these values are interpolated to the grid and the expected value computed in the response scale by applying the inverse of the link function. As this is done for each sample (from the posterior distribution), this approach can be used to compute any quantity of interest, such as, for example, the mean and standard error.

The first step is to build the data stack for the prediction on the mesh:

```
stk.mesh <- inla.stack(
  data = list(y = NA),
  A = list(1, 1),
  effects = list(s = 1:spde$n.spde,
    data.frame(Intercept = 1, oceanDist = oceanDist.mesh)),
```

```
   tag = 'mesh')

stk.b <- inla.stack(stk.dat, stk.mesh)
```

Next, the model is fitted again, and similar options to reduce the computations as in the previous example are set. In this case, an extra option is set to include in the output the precision matrix for each hyperparameter configuration for sampling from the joint posterior. This option is `control.compute = list(config = TRUE)`.

Hence, the code to fit the model is now:

```
rm.oceanD.1 <- inla(f.oceanD.1, family = 'Gamma',
  data = inla.stack.data(stk.b),
  control.predictor = list(A = inla.stack.A(stk.b),
    compute = TRUE, link = 1),
  quantiles = NULL,
  control.results = list(return.marginals.random = FALSE,
    return.marginals.predictor = FALSE),
  control.compute = list(config = TRUE)) # Needed to sample
```

Sampling from the posterior distribution using the fitted model is done using function `inla.posterior.sample()`. In the next example 1000 samples from the posterior will be obtained:

```
sampl <- inla.posterior.sample(n = 1000, result = rm.oceanD.1)
```

We do need to collect the index for the linear predictor corresponding to the stack data of the prediction scenario and use it to extract the corresponding elements of the latent field of each sample.

```
id.prd.mesh <- inla.stack.index(stk.b, 'mesh')$data
pred.nodes <- exp(sapply(sampl, function(x)
  x$latent[id.prd.mesh]))
```

Note that this is a matrix with the following dimensions:

```
dim(pred.nodes)
## [1]  967 1000
```

Computing the mean and standard deviation over the samples and projecting it is now:

```
sd.prd.s <- matrix(NA, nxy[1], nxy[2])
m.prd.s <- matrix(NA, nxy[1], nxy[2])

m.prd.s[xy.in] <- drop(Aprd %*% rowMeans(pred.nodes))
sd.prd.s[xy.in] <- drop(Aprd %*% apply(pred.nodes, 1, sd))
```

Finally, it is possible to check that this approach produces similar values as in the previous example (based on the computation of the posterior marginal):

```
cor(as.vector(m.prd.s), as.vector(m.prd), use = 'p')
## [1] 0.9998
cor(log(as.vector(sd.prd.s)), log(as.vector(sd.prd)), use = 'p')
## [1] 0.961
```

3

More than one likelihood

Chapters 1 and 2 have focused on describing the INLA and SPDE methodologies, respectively. In this chapter we introduce more complex models that require the use of several likelihoods. These examples will focus on models with a multivariate response, how to deal with measurement error and using part of the linear predictor of one variable as part of the linear predictor of another variable.

3.1 Coregionalization model

3.1.1 Motivation

In this section we present a way to fit the Bayesian coregionalization model similar to the ones proposed by Schmidt and Gelfand (2003) and Gelfand et al. (2002). A related code example to the one we present here can be found in Chapter 8 in Blangiardo and Cameletti (2015). These models are often used when measurement stations record several variables; for example, a station measuring pollution may register values of CO2, CO and NO2. Instead of modeling these as several univariate datasets, the models we present in this section deal with the joint dependency structure. Dependencies among the different outcomes are modeled through shared components at the predictor level.

Usually, in coregionalization models, the different responses are assumed to be observed at the same locations. With the INLA-SPDE approach, we do not require the different outcome variables to be measured at the same locations. Hence, in the code example below we show how to model responses observed at different locations. In this section we present a spatial model, and in Section 8.1 we generalize it to the space-time setting.

3.1.2 The model and parametrization

The case of three outcomes is defined considering the following equations:

$$y_1(\mathbf{s}) = \alpha_1 + z_1(\mathbf{s}) + e_1(\mathbf{s})$$
$$y_2(\mathbf{s}) = \alpha_2 + \lambda_1 z_1(\mathbf{s}) + z_2(\mathbf{s}) + e_2(\mathbf{s})$$
$$y_3(\mathbf{s}) = \alpha_3 + \lambda_2 z_1(\mathbf{s}) + \lambda_3 z_2(\mathbf{s}) + z_3(\mathbf{s}) + e_3(\mathbf{s}),$$

where the α_k are intercepts, $z_k(\mathbf{s})$ are spatial effects, λ_k are weights for some of the spatial effects and $e_k(\mathbf{s})$ are uncorrelated error terms, with $k = 1, 2, 3$.

This model can be fitted with INLA using the copy feature, which is explained in more detail in Section 1.6.2 and Section 3.3. In INLA, there is only one single linear predictor (one single formula). Thus, to implement the above three equations, we have to stack the linear predictors together into one long vector. Because of this, we have to copy the spatial effects $z_1(\mathbf{s})$ and $z_2(\mathbf{s})$ to represent the different contributions to the different parts of the formula.

3.1.3   Data simulation

The following parameters are used to simulate data from the model presented above:

```
# Intercept on reparametrized model
alpha <- c(-5, 3, 10)
# Random field marginal variances
m.var <- c(0.5, 0.4, 0.3)
# GRF range parameters:
range <- c(4, 3, 2)
# Copy parameters: reparameterization of coregionalization
# parameters
beta <- c(0.7, 0.5, -0.5)
# Standard deviations of error terms
e.sd <- c(0.3, 0.2, 0.15)
```

Similarly, the number of observations of each response variable is defined as follows:

```
n1 <- 99
n2 <- 100
n3 <- 101
```

In this example, we use a different number of observations for each response variable, and they are observed at different locations. In typical applications of coregionalization, however, all response variables will be measured at the same locations The location points are sampled at random on a $(0, 10) \times (0, 5)$ rectangular domain.

```
loc1 <- cbind(runif(n1) * 10, runif(n1) * 5)
loc2 <- cbind(runif(n2) * 10, runif(n2) * 5)
loc3 <- cbind(runif(n3) * 10, runif(n3) * 5)
```

The `book.rMatern()` function described in Section 2.1.3 will be used to simulate independent random field realizations for each time. The three random fields are simulated as follows. Note that we need, for example, the locations `loc2` for the field `u1` because this field is also used for the second variable.

```
set.seed(05101980)
z1 <- book.rMatern(1, rbind(loc1, loc2, loc3), range = range[1],
  sigma = sqrt(m.var[1]))
z2 <- book.rMatern(1, rbind(loc2, loc3), range = range[2],
  sigma = sqrt(m.var[2]))
z3 <- book.rMatern(1, loc3, range = range[3],
  sigma = sqrt(m.var[3]))
```

Finally, we obtain samples from the observations:

```
set.seed(08011952)

y1 <- alpha[1] + z1[1:n1] + rnorm(n1, 0, e.sd[1])
y2 <- alpha[2] + beta[1] * z1[n1 + 1:n2] + z2[1:n2] +
  rnorm(n2, 0, e.sd[2])
y3 <- alpha[3] + beta[2] * z1[n1 + n2 + 1:n3] +
  beta[3] * z2[n2 + 1:n3] + z3 + rnorm(n3, 0, e.sd[3])
```

3.1.4 Model fitting

This model only requires one mesh to fit all of the three spatial random fields. This makes it easier to link the different effects across different outcomes at different spatial locations. We choose to use all the locations when creating the mesh, as follows:

```
mesh <- inla.mesh.2d(rbind(loc1, loc2, loc3),
  max.edge = c(0.5, 1.25), offset = c(0.1, 1.5), cutoff = 0.1)
```

The next step is defining the SPDE model considering the PC-prior derived in Fuglstad et al. (2018) and described in Sections 1.6.5 and 2.3:

```
spde <- inla.spde2.pcmatern(
  mesh = mesh,
  prior.range = c(0.5, 0.01), # P(range < 0.5) = 0.01
  prior.sigma = c(1, 0.01)) # P(sigma > 1) = 0.01
```

For each of the parameters (i.e., coefficients) of the copied effects, the prior is Gaussian with zero mean and precision 10. These are defined as follows:

```
hyper <- list(theta = list(prior = 'normal', param = c(0, 10)))
```

The formula including all the terms in the model is defined as follows:

```
form <- y ~ 0 + intercept1 + intercept2 + intercept3 +
  f(s1, model = spde) + f(s2, model = spde) +
  f(s3, model = spde) +
  f(s12, copy = "s1", fixed = FALSE, hyper = hyper) +
  f(s13, copy = "s1", fixed = FALSE, hyper = hyper) +
  f(s23, copy = "s2", fixed = FALSE, hyper = hyper)
```

Similarly, the projection matrices for each set of locations are obtained as:

```
A1 <- inla.spde.make.A(mesh, loc1)
A2 <- inla.spde.make.A(mesh, loc2)
A3 <- inla.spde.make.A(mesh, loc3)
```

We organize the data by defining three stacks and then joining them:

```
stack1 <- inla.stack(
  data = list(y = cbind(as.vector(y1), NA, NA)),
  A = list(A1),
  effects = list(list(intercept1 = 1, s1 = 1:spde$n.spde)))

stack2 <- inla.stack(
  data = list(y = cbind(NA, as.vector(y2), NA)),
  A = list(A2),
  effects = list(list(intercept2 = 1, s2 = 1:spde$n.spde,
    s12 = 1:spde$n.spde)))

stack3 <- inla.stack(
  data = list(y = cbind(NA, NA, as.vector(y3))),
  A = list(A3),
```

```
effects = list(list(intercept3 = 1, s3 = 1:spde$n.spde,
   s13 = 1:spde$n.spde, s23 = 1:spde$n.spde)))

stack <- inla.stack(stack1, stack2, stack3)
```

We use a PC prior for the error precision (see Section 1.6.5):

```
hyper.eps <- list(hyper = list(theta = list(prior = 'pc.prec',
  param = c(1, 0.01))))
```

In this model there are 12 hyperparameters in total; two hyperparameters for each of the three spatial effects, one for each likelihood, and three copy parameters. To make the optimization process fast, the parameter values used in the simulation (plus some random noise) will be set as the initial values:

```
theta.ini <- c(log(1 / e.sd^2),
  c(log(range),
    log(sqrt(m.var)))[c(1, 4, 2, 5, 3, 6)], beta)
# We jitter the starting values to avoid artificially recovering
# the true values
theta.ini = theta.ini + rnorm(length(theta.ini), 0, 0.1)
```

Given that this model is complex and may take a long time to run, the empirical Bayes approach will be used, by setting `int.strategy = 'eb'` below, instead of integrating over the space of hyperparameters. The model is fitted with the following R code:

```
result <- inla(form, rep('gaussian', 3),
  data = inla.stack.data(stack),
  control.family = list(hyper.eps, hyper.eps, hyper.eps),
  control.predictor = list(A = inla.stack.A(stack)),
  control.mode = list(theta = theta.ini, restart = TRUE),
  control.inla = list(int.strategy = 'eb'))
```

We highlight the computational time below (measured in seconds).

```
##     Pre Running    Post    Total
##  6.1468 46.4021  0.3639 52.9128
```

The posterior mode of the model hyperparameters is displayed in Table 3.1, using `Mode = result$mode$theta`. We present this table because it shows the INLA internal representation of the model parameters.

TABLE 3.1: Posterior modes of some of the model parameters.

Parameter	Mode
$\log(1/\sigma_1^2)$	2.3269
$\log(1/\sigma_2^2)$	3.8249
$\log(1/\sigma_3^2)$	3.2441
$\log(\text{Range})$ for s_1	1.0511
$\log(\text{St. Dev.})$ for s_1	-0.5939
$\log(\text{Range})$ for s_2	0.9129
$\log(\text{St. Dev.})$ for s_2	-0.5561
$\log(\text{Range})$ for s_3	0.6893
$\log(\text{St. Dev.})$ for s_3	-0.8321
β_1	0.4695
β_2	0.6118
β_3	-0.3380

We can convert the posterior marginal distributions for the likelihood precision to standard deviation with:

```
p.sd <- lapply(result$internal.marginals.hyperpar[1:3],
  function(m) {
    inla.tmarginal(function(x) 1 / sqrt(exp(x)), m)
  })
```

The posterior marginal densities of the model hyper-parameters are summarized in Table 3.2. Estimates have been put together using the following code:

```
# Intercepts
tabcrp1 <- cbind(true = alpha, result$summary.fixed[, c(1:3, 5)])
# Precision of the errors
tabcrp2 <- cbind(
  true = c(e = e.sd),
  t(sapply(p.sd, function(m)
    unlist(inla.zmarginal(m, silent = TRUE))[c(1:3, 7)])))
colnames(tabcrp2) <- colnames(tabcrp1)
# Copy parameters
tabcrp3 <- cbind(
  true = beta, result$summary.hyperpar[10:12, c(1:3, 5)])
# The complete table
tabcrp <- rbind(tabcrp1, tabcrp2, tabcrp3)
```

TABLE 3.2: Summary of the posterior distributions of some of the parameters in the model.

Parameter	True	Mean	St. Dev.	2.5% quant.	97.5% quant.
α_1	-5.00	-4.3880	0.2173	-4.8146	-3.9618
α_2	3.00	2.8384	0.2219	2.4027	3.2738
α_3	10.00	10.3557	0.1904	9.9818	10.7292
σ_1	0.30	0.3145	0.0388	0.2449	0.3971
σ_2	0.20	0.1482	0.0298	0.0970	0.2137
σ_3	0.15	0.1907	0.0434	0.1145	0.2835
λ_1	0.70	0.4724	0.1926	0.0961	0.8528
λ_2	0.50	0.6254	0.1712	0.2949	0.9685
λ_3	-0.50	-0.3439	0.1663	-0.6742	-0.0193

The posterior marginals of the range and the standard deviations for each field are summarized in Table 3.3.

TABLE 3.3: Summary of the posterior distributions of some of the parameters of the spatial fields in the model.

Parameter	True	Mean	St. Dev.	2.5% quant.	97.5% quant.
Range for s1	4.0000	3.0124	0.8999	1.6356	5.1435
Range for s2	3.0000	2.5831	0.5525	1.6824	3.8430
Range for s3	2.0000	1.9873	0.6165	1.0062	3.4006
Stdev for s1	0.7071	0.5616	0.0943	0.4002	0.7697
Stdev for s2	0.6325	0.5873	0.0938	0.4275	0.7953
Stdev for s3	0.5477	0.4403	0.0853	0.2948	0.6295

The posterior mean of each random field can be projected to the data locations. Figure 3.1 compares the fitted values to the simulated ones and it seems that the model gives good estimates of the actual values. This figure has been produced using the following code:

```
par(mfrow = c(2, 3), mar = c(2.5, 2.5, 1.5, 0.5),
  mgp = c(1.5, 0.5, 0))
plot(drop(A1 %*% result$summary.random$s1$mean), z1[1:n1],
  xlab = 'Posterior mean', ylab = 'Simulated', asp = 1,
  main = 'z1 in y1')
abline(0:1)

plot(drop(A2 %*% result$summary.random$s1$mean), z1[n1 + 1:n2],
  xlab = 'Posterior mean', ylab = 'Simulated',
```

```
  asp = 1, main = 'z1 in y2')
abline(0:1)

plot(drop(A3 %*% result$summary.random$s1$mean),
  z1[n1 + n2 + 1:n3],
  xlab = 'Posterior mean', ylab = 'Simulated',
  asp = 1, main = 'z1 in y3')
abline(0:1)

plot(drop(A2 %*% result$summary.random$s2$mean), z2[1:n2],
  xlab = 'Posterior mean', ylab = 'Simulated',
  asp = 1, main = 'z2 in y2')
abline(0:1)

plot(drop(A3 %*% result$summary.random$s2$mean), z2[n2 + 1:n3],
  xlab = 'Posterior mean', ylab = 'Simulated',
  asp = 1, main = 'z2 in y3')
abline(0:1)

plot(drop(A3 %*% result$summary.random$s3$mean), z3[1:n3],
  xlab = 'Posterior mean', ylab = 'Simulated',
  asp = 1, main = 'z3 in y3')
abline(0:1)
```

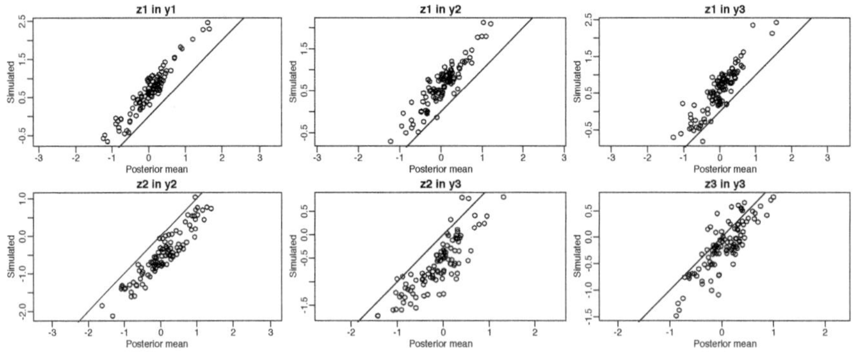

FIGURE 3.1 Simulated versus posterior mean fitted for the spatial fields.

3.2 Joint modeling: Measurement error model

In this section we challenge the assumption that the covariate is measured accurately. Specifically, we work with measurement error models that account for the uncertainty of covariate measurements. Introducing uncertainty in the covariate measurements is not common, not because we think covariates are perfectly measured, but because making models for measurement error is hard. In these models we have an outcome that varies over space, $y(\mathbf{s})$, that depends on a covariate that also varies over space $x(\mathbf{s})$. This $x(\mathbf{s})$ is assumed to be observed with error. Further, we allow for spatial misalignment between the response $\mathbf{y}$ and the covariate values $\mathbf{x}$. A related example can be found in Chapter 8 of Blangiardo and Cameletti (2015).

We consider a response variable that is observed at a set of n_y locations, represented as grey dots in Figure 3.2. The covariate $\mathbf{x}$ is observed at a set of n_x locations, shown as triangles in Figure 3.2. In some cases, y_i and x_i could be at the same location. Locations shown in Figure 3.2 are simulated by:

```
n.x <- 70
n.y <- 50
set.seed(1)
loc.y <- cbind(runif(n.y) * 10, runif(n.y) * 5)
loc.x <- cbind(runif(n.x) * 10, runif(n.x) * 5)

n.x <- nrow(loc.x)
n.y <- nrow(loc.y)
```

In this section we implement the classical measurement error (MEC, Muff et al., 2014), where we observe a proxy $\mathbf{w}$ for $\mathbf{x}$. Specifically, $\mathbf{w} = \mathbf{x} + \epsilon$ where the error ϵ is considered independent of $\mathbf{x}$. Alternatively, we could assume that the error ϵ is independent of $\mathbf{w}$; this is the so-called Berkson measurement error (Muff et al., 2014).

The linear predictor for the outcome $\mathbf{y}$ is modeled as

$$
\begin{aligned}
\eta_{\mathbf{y}} &= \alpha_y + \beta \mathbf{A}_y \mathbf{x}(\mathbf{s}) + \mathbf{A}_y \mathbf{v}(\mathbf{s}) \\
\mathbf{w} &= \mathbf{A}_x \mathbf{x}(\mathbf{s}) + \epsilon \\
\mathbf{x}(\mathbf{s}) &= \alpha_x + \mathbf{m}(\mathbf{s}),
\end{aligned}
\tag{3.1}
$$

where α_y and α_x are intercept parameters for $\mathbf{y}$ and $\mathbf{x}$, respectively, β is the regression coefficient of $\mathbf{x}$ on $\mathbf{y}$, ϵ is a Gaussian noise with variance σ_ϵ^2, $\epsilon \sim N(0, \sigma_\epsilon^2 \mathbf{I})$, and both $\mathbf{v}(\mathbf{s})$ and $\mathbf{m}(\mathbf{s})$ are considered to be spatially structured, which implies that $\mathbf{x}$ is also spatially structured through $\mathbf{m}(\mathbf{s})$. This allows us to have $\mathbf{y}$ and $\mathbf{x}$ being collected at different locations. Thus, we have the

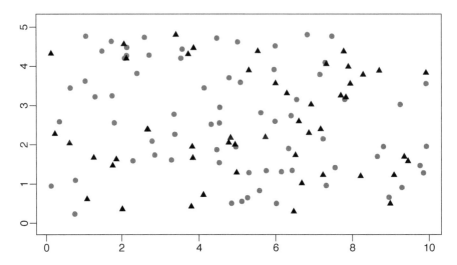

FIGURE 3.2 Locations for the covariate (gray dots) and outcome (black triangles).

projection matrices $\mathbf{A}_x$ and $\mathbf{A}_y$ to project any of the spatial processes at the **x** and **y** locations, respectively.

This setup is independent of the choice of the likelihood for **y**. In our example, we consider the Poisson likelihood, i.e.

$$\mathbf{y} \sim \text{Poisson}\left(e^{\eta_y}\right). \tag{3.2}$$

3.2.1 Simulation from the model

To simulate from our model, we first simulate **v** and **m** using the `book.rMatern` function defined in Section 2.1.3. Field **m** is simulated at both sets of locations in order to be able to simulate **y** later.

```
range.v <- 3
sigma.v <- 0.5
range.m <- 3
sigma.m <- 1
set.seed(2)
v <- book.rMatern(n = 1, coords = loc.y, range = range.v,
  sigma = sigma.v, nu = 1)
m <- book.rMatern(n = 1, coords = rbind(loc.x, loc.y),
  range = range.m, sigma = sigma.m, nu = 1)
```

Next, the remaining parameters are set:

```
alpha.y <- 2
alpha.x <- 5
beta.x <- 0.3
sigma.e <- 0.2
```

We now simulate **x** and **w**, at the **x** and **y** locations. We compare **x** and **w** in Figure 3.3.

```
x.x <- alpha.x + m[1:n.x]
w.x <- x.x + rnorm(n.x, 0, sigma.e)
x.y <- alpha.x + m[n.x + 1:n.y]
w.y <- x.y + rnorm(n.y, 0, sigma.e)
```

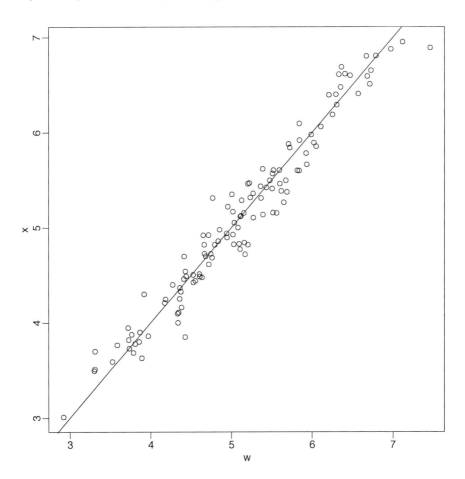

FIGURE 3.3 Simulated values of **x** versus **w**.

Outcome **y** is simulated as follows:

```
eta.y <- alpha.y + beta.x * x.y + v
set.seed(3)
yy <- rpois(n.y, exp(eta.y))
```

3.2.2   Model fitting

We build the mesh taking into account the value chosen for the ranges, `range.v` and `range.m`, of the spatial processes.

```
mesh <- inla.mesh.2d(rbind(loc.x, loc.y),
   max.edge = min(range.m, range.v) * c(1 / 3, 1),
   offset = min(range.m, range.v) * c(0.5, 1.5))
```

This mesh is made of 479 points. The same mesh will be considered to build the SPDE model for **v** and **m**.

The projection matrices for the covariate and outcome locations are defined as follows:

```
Ax <- inla.spde.make.A(mesh, loc = loc.x)
Ay <- inla.spde.make.A(mesh, loc = loc.y)
```

For the parameters of the SPDE model, namely the practical range and the marginal standard deviation, we consider the PC-prior derived in Fuglstad et al. (2018), defined as:

```
spde <- inla.spde2.pcmatern(mesh = mesh,
   prior.range = c(0.5, 0.01), # P(practic.range < 0.5) = 0.01
   prior.sigma = c(1, 0.01)) # P(sigma > 1) = 0.01
```

In practice, the available data is w_j, $j = 1, ..., n_x$ and y_i, $i = n_x + 1, ..., n_x + n_y$. Thus, we need a model for **x** that is able to compute (or predict) it at the outcome locations y_i. From (3.1) we can write

$$0 = \alpha_x + \mathbf{m} - \mathbf{x}. \tag{3.3}$$

Because **m** is a GF, it can be defined over the entire area and used to predict **x** at any point, particularly at the **y** locations. This negative **x** term is then copied into the **y** linear predictor in order to estimate β, using the `copy` feature.

To construct the stack for this model, we build one stack for each likelihood and then join them together. The data are supplied as a three column matrix where each column matches each likelihood in the `inla()` call. The first column contains the n_y "faked zeros", while the other two columns contain NA. The effects should contain the intercept α_x, the index set for **m** and an index from 1 to n_y to compute the $-\mathbf{x}$ term at the **y** locations. Since, α_x and $-\mathbf{x}$ are associated with each element of the "faked zero" observations, we group them in the same list element of the data stack. The index set for **m** is another element on the effects list. These effects are associated respectively with the identity projector (or just a single number one) and the projector built earlier for the **y** locations. As we will build more than one data stack, we use the `tag` to keep track of it. All of this can be written as follows:

```
stk.0 <- inla.stack(
  data = list(Y = cbind(rep(0, n.y), NA, NA)),
  A = list(1, Ay),
  effects = list(data.frame(alpha.x = 1, x0 = 1:n.y, x0w = -1),
    m = 1:spde$n.spde),
  tag = 'dat.0')
```

We now consider the **w** data in the second column. We have that

$$\mathbf{w} \sim N(\mathbf{x} = \alpha_x + \mathbf{m}, \sigma_\epsilon^2 \mathbf{I}), \tag{3.4}$$

giving:

```
stk.x <- inla.stack(
  data = list(Y = cbind(NA, w.x, NA)),
  A = list(1, Ax),
  effects = list(alpha.x = rep(1, n.x), m = 1:mesh$n),
  tag = 'dat.x')
```

To build the data stack for **y** we need to include the terms for the intercept α_y, the index set for **x** which will be copied from the "faked zero" observations model and the index for the **v** term:

```
stk.y <- inla.stack(
  data = list(Y = cbind(NA, NA, yy)),
  A = list(1, Ay),
  effects = list(data.frame(alpha.y = 1, x = 1:n.y),
    v = 1:mesh$n),
  tag = 'dat.y')
```

We must supply only one formula for `inla()`, which should contain all the model terms. The terms in the right hand side of the formula that are not present in the data stack will just be ignored in the linear predictor for the corresponding data observations. The regression coefficient β is fitted with the copy feature, assuming a zero-mean Gaussian prior with precision 1. For the "faked zero" observations we assume for α_x the default prior which is $N(0, 1000)$. The SPDE model for **m** is already defined, and **x** gets an iid (independent and identically distributed) Gaussian random effect with low fixed precision. The **x** is forced to be equal to $\alpha_x - \mathbf{m}$ by considering a Gaussian likelihood with some (fixed) high precision and "faked zero" data.

```
form <- Y ~  0 + alpha.x + alpha.y +
  f(m, model = spde) + f(v, model = spde) +
  f(x0, x0w, model = 'iid',
    hyper = list(theta = list(initial = -20, fixed = TRUE))) +
  f(x, copy = 'x0', fixed = FALSE,
    hyper = list(theta = list(prior='normal', param = c(0, 1)))))
hfix <- list(hyper = list(theta = list(initial = 20,
  fixed = TRUE)))
```

We consider a PC-prior for σ_ϵ assuming $P(\sigma_\epsilon < 0.2) = 0.5$:

```
pprec <- list(hyper = list(theta = list(prior = 'pc.prec',
  param=c(0.2, 0.5)))))
```

The model is then fitted stacking all the data previously defined:

```
stk <- inla.stack(stk.0, stk.x, stk.y)
res <- inla(form, data = inla.stack.data(stk),
  family = c('gaussian', 'gaussian', 'poisson'),
  control.predictor = list(compute = TRUE,
    A = inla.stack.A(stk)),
  control.family = list(hfix, pprec, list()))
```

3.2.3 Results

Table 3.4 shows the true values of model parameters used in the simulation as well as summaries of their posterior distributions. Figure 3.4 shows the posterior distribution of the regression parameters. We observe that for most of the cases, the true values of the parameters fall within the areas of high probability of the corresponding posterior distributions. The posterior distributions of the parameters of each random field are shown in Figure 3.5. The true values are also in the areas of high probability of their corresponding posterior marginal distributions.

TABLE 3.4: Summary of the posterior distributions of the parameters in the model.

Parameter	True	Mean	St. Dev.	2.5% quant.	97.5% quant.
α_x	5.0	5.1465	0.3781	4.3721	5.8908
α_y	2.0	2.3034	0.5320	1.2254	3.3278
σ_ϵ	0.2	0.1879	0.0571	0.0947	0.3167
Range for m	3.0	2.8993	0.7056	1.8315	4.5765
Stdev for m	1.0	1.0042	0.1589	0.7432	1.3655
Range for v	3.0	4.1037	1.6281	2.0129	8.2359
Stdev for v	0.5	0.4765	0.1008	0.3168	0.7114
Beta for x	0.3	0.2573	0.0978	0.0692	0.4544

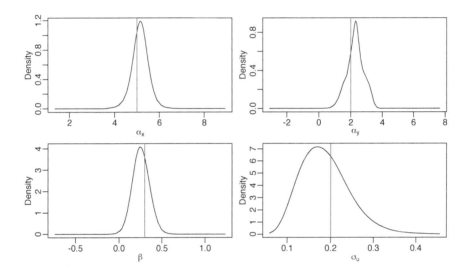

FIGURE 3.4 Posterior distribution of the intercepts, the regression coefficient and σ_ϵ. Vertical lines represent the actual value of the parameter used in the simulations.

Another interesting result is the prediction of the covariate at the response locations. Given that **m** was simulated, we can assess how good the predictions for **m** are at the **x** and **y** locations. We can project the posterior mean and standard deviations of **m** at all the n_x and n_y locations:

```
mesh2locs <- rbind(Ax, Ay)
m.mu <- drop(mesh2locs %*% res$summary.ran$m$mean)
m.sd <- drop(mesh2locs %*% res$summary.ran$m$sd)
```

Then, we can approximate the 95% credible intervals for **m** at each location,

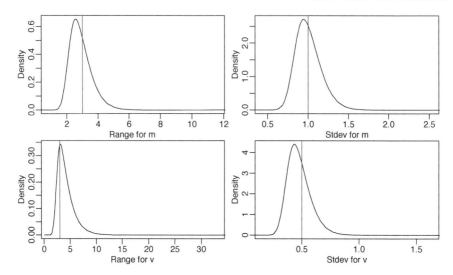

FIGURE 3.5 Posterior marginal distributions of the hyperparameters for both random fields. Vertical lines represent the actual value of the parameter used in the simulations.

assuming a posterior normal distribution. This is shown in Figure 3.6, where the blue line represents the situation where the predicted value is equal to the simulated value.

We visualize the expected value of **m** and **v** in Figure 3.7 with the following code:

```
# Create grid for projection
prj <- inla.mesh.projector(mesh, xlim = c(0, 10), ylim = c(0, 5))

# Settings for plotting device
par(mfrow = c(2, 1), mar = c(0.5, 0.5, 0.5, 0.5),
  mgp = c(1.5, 0.5, 0))

# Posterior mean of 'm'
book.plot.field(field = res$summary.ran$m$mean, projector = prj)
points(loc.x, cex = 0.3 + (m - min(m))/diff(range(m)))
# Posterior mean of 'v'
book.plot.field(field = res$summary.ran$v$mean, projector = prj)
points(loc.y, cex = 0.3 + (v - min(v))/diff(range(v)))
```

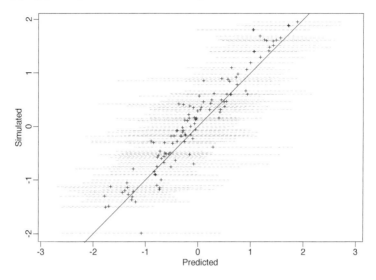

FIGURE 3.6 Simulated versus posterior mean of **m** (+) and the approximated 95% credible intervals (grey dashed lines). The blue line represents the situation when the posterior mean is equal to the simulated values.

3.3 Copying part of or the entire linear predictor

In this section we provide more insight into the technique of copying part of a linear predictor, which was used in both previous sections. In general, this is needed for all joint modeling.

Assume that data $y_1(s)$, $y_2(s)$ and $y_3(s)$ have been collected at location s. Also, consider the following models for the three types of observation:

$$
\begin{aligned}
y_1(s) &= \beta_0 + \beta_1 x(s) + A(s, s_0)b(s_0) + \epsilon_1(s) & (3.5)\\
y_2(s) &= \beta_2(\beta_0 + \beta_1 x(s)) + \epsilon_2(s) & (3.6)\\
y_3(s) &= \beta_3(\beta_0 + \beta_1 x(s) + A(s, s_0)b(s_0)) + \epsilon_3(s). & (3.7)
\end{aligned}
$$

Here, the SPDE model is defined at the mesh nodes $b(s_0)$ where $A(s, s_0)$ is the projection matrix, and ϵ_j, $j = 1, 2, 3$, are observation errors considered as zero mean Gaussians with variances σ_j^2. In this setting, there is a different linear model for each outcome. Also, there is a common effect that is scaled from one linear predictor into another, where β_2 and β_3 are the scaling parameters.

We define the following model terms:

- $\eta_0(s) = \beta_0 + \beta_1 x(s)$

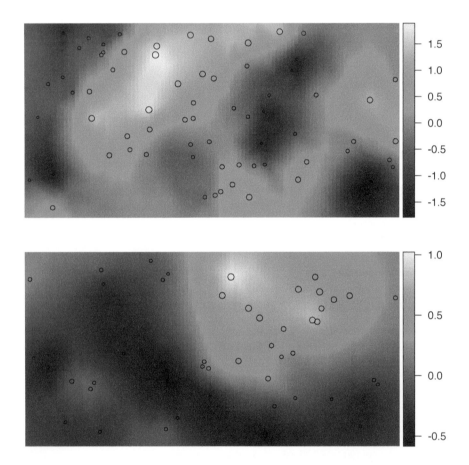

FIGURE 3.7 Posterior mean for **m** and the **x** locations plotted as points with size proportional to the simulated values for **m** (top). Posterior mean for **v** and the **y** locations plotted as points with size proportional to the simulated values for **v** (bottom).

- $\eta_1(s) = \eta_0(s) + A(s, s_0)b(s_0)$
- $\eta_2(s) = \beta_2\eta_0(s)$
- $\eta_3(s) = \beta_3\eta_1(s)$

We will show how to copy η_0 into η_2, and η_1 into η_3, in order to estimate β_2 and β_3. Furthermore, all the three observation vectors, y_1, y_2 and y_3, are assumed to be observed at the same locations.

3.3.1 Generating the data

The set of parameters in the model for β_j, $j = 0, 1, 2, 3$, is defined as follows:

```
beta0 = 5
beta1 = 1
beta2 = 0.5
beta3 = 2
```

Then, the standard deviations of the error terms are defined as:

```
s123 <- c(0.1, 0.05, 0.15)
```

For the $b(s)$ process, a Matérn covariance function with κ_b, σ_b^2 and $\nu = 1$ (fixed) is considered, so that the following parameters are defined:

```
kappab <- 10
sigma2b <- 1
```

To obtain a realization of the spatial process, a set of locations is considered as follows:

```
n <- 50
loc <- cbind(runif(n), runif(n))
```

As in previous sections, the sample from the Matérn process will be obtained using function `book.rMatern` (see Section 2.1.3).

```
b <- book.rMatern(n = 1, coords = loc,
  range = sqrt(8) / 10, sigma = 1)
```

We plot this sample in Figure 3.8. We simulate a covariate as follows:

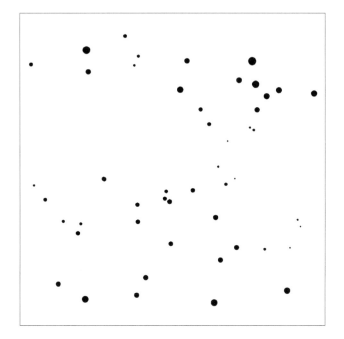

FIGURE 3.8 Locations of the simulated dataset. Point size is proportional to the value of the underlying Matérn process.

```
x <- sqrt(3) * runif(n, -1, 1)
```

Then, the required linear predictors are computed:

```
eta1 <- beta0 + beta1 * x + b
eta2 <- beta2 * (beta0 + beta1 * x)
eta3 <- beta3 * eta1
```

Finally, the observations are obtained:

```
y1 <- rnorm(n, eta1, s123[1])
y2 <- rnorm(n, eta2, s123[2])
y3 <- rnorm(n, eta3, s123[3])
```

3.3.2   The fake zero trick for fitting this model

There is more than one way to fit this model using the INLA package. The main point is the need to compute $\boldsymbol{\eta}_0(\boldsymbol{s}) = \beta_0 + \beta_1 \boldsymbol{x}(\boldsymbol{s})$ and $\boldsymbol{\eta}_1(\boldsymbol{s}) = \beta_0 +$

$\beta_1 \boldsymbol{x}(\boldsymbol{s}) + \boldsymbol{A}(\boldsymbol{s}, \boldsymbol{s}_0)\boldsymbol{b}(\boldsymbol{s})$ from the first observation equation in order to copy it to the second and third observation equations. So, a model that computes $\boldsymbol{\eta}_0(\boldsymbol{s})$ and $\boldsymbol{\eta}_1(\boldsymbol{s})$ explicitly needs to be defined.

The way we choose is to minimize the size of the graph generated by the model (Rue et al., 2017). First, the following equations are considered:

$$\boldsymbol{0}(\boldsymbol{s}) \;\;=\;\; \boldsymbol{A}(\boldsymbol{s}, \boldsymbol{s}_0)\boldsymbol{b}(\boldsymbol{s}_0) + \boldsymbol{\eta}_0(\boldsymbol{s}) + \boldsymbol{\epsilon}_1(\boldsymbol{s}) - \boldsymbol{y}_1(\boldsymbol{s}) \tag{3.8}$$
$$\boldsymbol{0}(\boldsymbol{s}) \;\;=\;\; \boldsymbol{\eta}_1(\boldsymbol{s}) + \boldsymbol{\epsilon}_1(\boldsymbol{s}) - \boldsymbol{y}_1(\boldsymbol{s}) \tag{3.9}$$

where only $\boldsymbol{A}(\boldsymbol{s}, \boldsymbol{s}_0)$, $\boldsymbol{x}(\boldsymbol{s})$ and $\boldsymbol{y}_1(\boldsymbol{s})$ are known. For the $\boldsymbol{\eta}_0(\boldsymbol{s})$ and $\boldsymbol{\eta}_2(\boldsymbol{s})$ terms, we assume independent and identically distributed (i.e., iid) models with low fixed precision. With this fixed high variance, each element in $\boldsymbol{\eta}_0(\boldsymbol{s})$ and $\boldsymbol{\eta}_1(\boldsymbol{s})$ can take any value.

However, these values will be forced to be $\beta_0 + \beta_1\boldsymbol{x}(\boldsymbol{s})$ and $\beta_0 + \beta_1\boldsymbol{x}(\boldsymbol{s}) + \boldsymbol{A}(\boldsymbol{s}, \boldsymbol{s}_0)\boldsymbol{b}(\boldsymbol{s}_0)$ by considering a Gaussian likelihood for the "faked zero" observations with a high fixed precision value (i.e., a low fixed likelihood variance). For details and examples of this approach see Section 3.2, and Ruiz-Cárdenas et al. (2012), Martins et al. (2013) and Chapter 8 in Blangiardo and Cameletti (2015).

Since three Gaussian likelihoods are considered for the observed data, three error terms can be included in the linear predictors and can fix the likelihood precisions to high values. With this setting, the three Gaussian likelihoods with high fixed precisions become a single Gaussian likelihood.

3.3.3 Fitting the model

In order to fit the $\boldsymbol{b}(\boldsymbol{s})$ term, we define a SPDE Matérn model. For this, the first step is to set a mesh:

```
mesh <- inla.mesh.2d(
  loc.domain = cbind(c(0, 1, 1, 0), c(0, 0, 1, 1)),
  max.edge = c(0.1, 0.3), offset = c(0.05, 0.35), cutoff = 0.05)
```

Next, the projection matrix is defined:

```
As <- inla.spde.make.A(mesh, loc)
```

Then, the SPDE model is defined:

```
spde <- inla.spde2.pcmatern(mesh, alpha = 2,
```

```
prior.range = c(0.05, 0.01),
prior.sigma = c(1, 0.01))
```

The data have to be organized using the inla.stack() function. The data
stack for the first observation vector is:

```
stack1 <- inla.stack(
  data = list(y = y1),
  A = list(1, As, 1),
  effects = list(
    data.frame(beta0 = 1, beta1 = x),
    s = 1:spde$n.spde,
    e1 = 1:n),
  tag = 'y1')
```

Here, the e1 term will be used to fit ϵ_1. Similarly, the data stack for the first
"faked zero" observations is:

```
stack01 <- inla.stack(
  data = list(y = rep(0, n), offset = -y1),
  A = list(As, 1),
  effects = list(s = 1:spde$n.spde,
    list(e1 = 1:n, eta1 = 1:n)),
  tag = 'eta1')
```

Here, the data stack includes minus the first observation vector as an offset.
The stack for the second "faked zero" observation is:

```
stack02 <- inla.stack(
  data = list(y = rep(0, n), offset = -y1),
  effects = list(list(e1 = 1:n, eta2 = 1:n)),
  A = list(1),
  tag = 'eta2')
```

The stack for the second vector of observations now considers an index set to
compute the η_1 term copied from the first "faked zero" observations:

```
stack2 <- inla.stack(
  data = list(y = y2),
  effects = list(list(eta1c = 1:n, e2 = 1:n)),
  A = list(1),
  tag = 'y2')
```

In a similar way, the third observation stack also includes an index set to compute the η_2 term copied from the second "faked zero" observations:

```
stack3 <- inla.stack(
  data = list(y = y3),
  effects = list(list(eta2c = 1:n, e3 = 1:n)),
  A = list(1),
  tag = 'y3')
```

Once all the different data stacks have been defined, they are combined into a new one that will be used to fit the model:

```
stack <- inla.stack(stack1, stack01, stack02, stack2, stack3)
```

We use the default priors for most of the parameters. For the three variance errors in the observations PC-priors will be set as follows:

```
pcprec <- list(theta = list(prior = 'pcprec',
  param = c(0.5, 0.1)))
```

```
formula123 <- y ~ 0 + beta0 + beta1 +
  f(s, model = spde) + f(e1, model = 'iid', hyper = pcprec) +
  f(eta1, model = 'iid',
    hyper = list(theta = list(initial = -10, fixed = TRUE))) +
  f(eta2, model = 'iid',
    hyper = list(theta = list(initial = -10, fixed = TRUE))) +
  f(eta1c, copy = 'eta1', fixed = FALSE) +
  f(e2, model = 'iid', hyper = pcprec) +
  f(eta2c, copy = 'eta2', fixed = FALSE) +
  f(e3, model = 'iid', hyper = pcprec)
```

Finally, the model is fitted:

```
res123 <- inla(formula123,
  data = inla.stack.data(stack),
  offset = offset,
  control.family = list(list(
    hyper = list(theta = list(initial = 10, fixed = TRUE)))),
  control.predictor = list(A = inla.stack.A(stack)))
```

3.3.4   Model results

A summary of the fixed effects β_0 and β_1, as well as the β_2 and β_3 parameters is available in Table 3.5.

TABLE 3.5: Posterior modes of some of the model parameters.

Parameter	True	Mean	St. Dev.	2.5% quant.	97.5% quant.
β_0	5.0	5.1555	0.1550	4.8810	5.3819
β_1	1.0	0.9805	0.0309	0.9195	1.0344
Beta for eta1c	0.5	0.4904	0.0151	0.4594	0.5186
Beta for eta2c	2.0	2.0099	0.0054	1.9996	2.0207

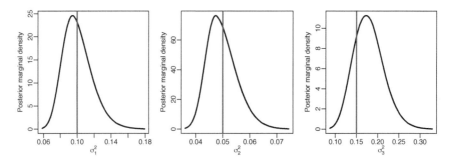

FIGURE 3.9 Observation error standard deviations. Vertical lines represent the actual values of the parameters.

Figure 3.9 shows the posterior marginal distribution of the standard deviation for ϵ_1, ϵ_2 and ϵ_3. Similarly, Figure 3.10 shows the posterior marginals of the random field parameters.

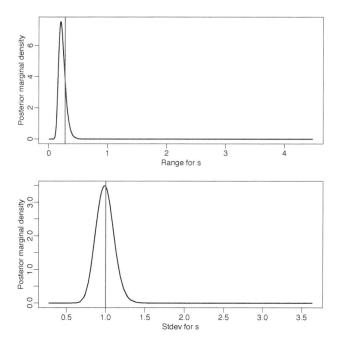

FIGURE 3.10 Posterior marginal distributions for the random field parameters. Vertical lines represent the actual values of the parameters.

4

Point processes and preferential sampling

4.1 Introduction

A point pattern records the occurrence of events in a study region. Typical examples include the locations of trees in a forest or the GPS coordinates of disease cases in a region. The locations of the observed events depend on an underlying spatial process, which is often modeled using an intensity function $\lambda(\mathbf{s})$. The intensity function measures the average number of events per unit of space, and it can be modeled to depend on covariates and other effects. See Diggle (2014) or Baddeley et al. (2015) for a recent summary on the analysis of spatial and spatio-temporal point patterns.

Under the log-Cox point process model assumption, we model the log intensity of the Cox process with a Gaussian linear predictor. In this case, the log-Cox process is known as a log-Gaussian Cox process (LGCP, Møller et al., 1998), and inference can be made using `INLA`. A Cox process is just a name for a Poisson process with varying intensity; thus we use the Poisson likelihood. The original approach that was used to fit these models in INLA (and other software) divides the study region into cells, which form a lattice, and counts the number of points in each one (Møller and Waagepetersen, 2003). These counts can be modeled using a Poisson likelihood conditional on a Gaussian linear predictor and `INLA` can be used to fit the model (Illian et al., 2012). This can be done with the techniques already shown in this book.

In this chapter we focus on a new approach considering SPDE models directly, developed in Simpson et al. (2016). This approach has a nice theoretical justification and considers a direct approximation of the log-Cox point process model likelihood. Observations are modeled considering its exact location instead of binning them into cells. Along with the flexibility for defining a mesh, this approach can handle non-rectangular areas without wasting computational effort on a large rectangular area.

4.1.1 Definition of the LGCP

The Cox process is a Poisson process with intensity $\lambda(\mathbf{s})$ that varies in space. Given some area A (for example a grid cell), the probability of observing a certain number of points in that area follows a Poisson distribution with intensity (expected value)

$$\lambda_A = \int_A \lambda(\mathbf{s}) \, d\mathbf{s}.$$

The log-Gaussian part of the LGCP name comes from modeling $\log(\lambda(\mathbf{s}))$ as a latent Gaussian (conditional on a set of hyper-parameters), in the typical GLM/GAM framework.

4.1.2   Data simulation

Data simulated here will be used later in Section 4.2 and Section 4.3. To sample from a log-Cox point process the function used is rLGCP() from the spatstat package (Baddeley et al., 2015). We use a $(0, 3) \times (0, 3)$ simulation window. In order to define this window using function owin() (in package spatstat):

```
library(spatstat)
win <- owin(c(0, 3), c(0, 3))
```

The rLGCP function uses the GaussRF() function, in package RandomFields (Schlather et al., 2015), to simulate from the spatial field over a grid. There is an internal parameter to control the resolution of the grid, which we specify to give 300 pixels in each direction:

```
npix <- 300
spatstat.options(npixel = npix)
```

We model the intensity as

$$\log(\lambda(\mathbf{s})) = \beta_0 + S(\mathbf{s}),$$

with β_0 a fixed value and $S(\mathbf{s})$ a Gaussian spatial process with Matérn covariance and zero mean. Parameter β_0 can be regarded as a global mean level for the log intensity; i.e. the log-intensity fluctuates about it according to the spatial process $S(\mathbf{s})$.

If there is no spatial field, the expected number of points is e^{β_0} times the area of the window. This means that the expected number of points is:

```
beta0 <- 3
exp(beta0) * diff(range(win$x)) * diff(range(win$y))
## [1] 180.8
```

Hence, this value of `beta0` will produce a reasonable number of points in the following simulations. If we set `beta0` too small, we will get almost no points, and we would not be able to produce reasonable results. It is also possible to use a function on several covariates, e.g. a GLM (see Section 4.2).

In this chapter we use a Matérn covariance function with $\nu = 1$. The other parameters are the variance and scale. The following values for these parameters will produce a smooth intensity of the point process:

```
sigma2x <- 0.2
range <- 1.2
nu <- 1
```

The value of `sigma2x` is set to make the log-intensity vary a bit around the mean, but always within a reasonable range of values. Furthermore, with these parameters $\nu = 1$ and the range of the spatial process $S(\mathbf{s})$ is (about) 2, which produces smooth changes in the current study window. Smaller values of the practical range will produce a spatial process $S(s)$ (and, in turn the intensity of the spatial process) that changes rapidly over the study window. Similarly, very large values of the practical range will produce an almost constant spatial process $S(s)$, so that the log-intensity will be very close to β_0 at all points of the study window.

The points of the point process are simulated as follows:

```
library(RandomFields)
set.seed(1)
lg.s <- rLGCP('matern', beta0, var = sigma2x,
  scale = range / sqrt(8), nu = nu, win = win)
```

Both the spatial field and the point pattern are returned. The coordinates of the observed events of the point pattern can be obtained as follows:

```
xy <- cbind(lg.s$x, lg.s$y)[, 2:1]
```

The number of simulated points is:

```
(n <- nrow(xy))
## [1] 189
```

The exponential of the simulated values of the spatial field are returned as the `Lambda` attribute of the object. Below, we extract the values of $\lambda(\mathbf{s})$ and summarize the $\log(\lambda(\mathbf{s}))$.

```
Lam <- attr(lg.s, 'Lambda')
rf.s <- log(Lam$v)
summary(as.vector(rf.s))
##    Min. 1st Qu.  Median    Mean 3rd Qu.    Max.
##    1.59    2.64    2.91    2.95    3.26    4.38
```

Figure 4.1 shows the simulated spatial field and the point pattern.

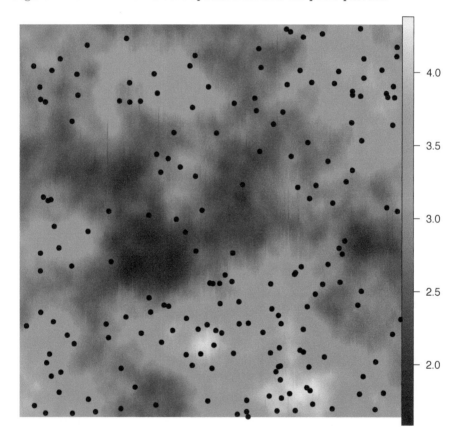

FIGURE 4.1 Simulated intensity of the point process and simulated point pattern (black dots).

4.1.3   Inference

Following Simpson et al. (2016), the parameters of the log-Gaussian Cox point process model can be estimated with INLA. In simplified terms, we will construct an augmented dataset and run a Poisson regression with INLA. The augmented data set is made of a binary response, with 1 for the observed points and 0 for some dummy observations. Both the observed and dummy observations will have associated 'expected values' or weights that will be included in the Poisson regression. This will be explained step by step in the following sections.

The mesh and the weights

For appropriate inference with the LGCP, we must take some care when building the mesh. In the particular case of the analysis of point patterns, we do not usually use the location points as mesh nodes. We need a mesh that covers the study region; for this we use the `loc.domain` to build the mesh. Further, we only use a small first outer extension, but no second outer extension.

```
loc.d <- 3 * cbind(c(0, 1, 1, 0, 0), c(0, 0, 1, 1, 0))
mesh <- inla.mesh.2d(loc.domain = loc.d, offset = c(0.3, 1),
  max.edge = c(0.3, 0.7), cutoff = 0.05)
nv <- mesh$n
```

This mesh can be seen in Figure 4.2.

The SPDE model will be defined considering the PC-priors derived in Fuglstad et al. (2018) for the model parameters range and marginal standard deviation. These are defined as follows:

```
spde <- inla.spde2.pcmatern(mesh = mesh,
  # PC-prior on range: P(practic.range < 0.05) = 0.01
  prior.range = c(0.05, 0.01),
  # PC-prior on sigma: P(sigma > 1) = 0.01
  prior.sigma = c(1, 0.01))
```

The SPDE approach for point pattern analysis defines the model at the nodes of the mesh. To fit the log-Cox point process model, these points are considered as integration points. The method in Simpson et al. (2016) defines the expected number of events to be proportional to the area around the node (the areas of the polygons in the dual mesh). This means that at the nodes of the mesh with larger triangles, there are also larger expected values. `inla.mesh.fem(mesh)$va` gives this value for every mesh node. These values for the nodes in the inner domain can be used to compute the intersection between the dual mesh polygons and the study domain polygon. To do that, we use the function `book.mesh.dual()`:

```
dmesh <- book.mesh.dual(mesh)
```

This function is available in the file `spde-book-functions.R`, and it returns

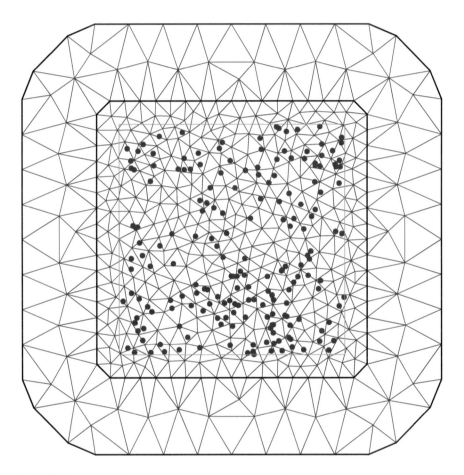

FIGURE 4.2 Mesh used to fit a log-Gaussian Cox process to a point pattern.

the dual mesh in an object of class `SpatialPolygons`. We have plotted the dual mesh in Figure 4.3.

The domain polygon can be converted into a `SpatialPolygons` class as follows:

```
domain.polys <- Polygons(list(Polygon(loc.d)), '0')
domainSP <- SpatialPolygons(list(domain.polys))
```

Because the mesh is larger than the study area, we need to compute the intersection between each polygon in the dual mesh and the study area:

```
library(rgeos)
w <- sapply(1:length(dmesh), function(i) {
  if (gIntersects(dmesh[i, ], domainSP))
    return(gArea(gIntersection(dmesh[i, ], domainSP)))
  else return(0)
})
```

The sum of these weights is the area of the study region:

```
sum(w)
## [1] 9
```

We can also check how many integration points have zero weight:

```
table(w > 0)
##
## FALSE   TRUE
##   198    294
```

The integration points with zero weight are identified in red in Figure 4.3. Note how all of them, and the associated polygons, are outside the study region.

Data and projection matrices

The vector of weights we have computed is exactly what we need to use as the exposure (E) in the Poisson likelihood in `INLA` (with the minor modification that $\log(E)$ is defined as zero if $E = 0$). We augment the vector of ones for the observations (representing the points) with a sequence of zeros (representing the mesh nodes):

```
y.pp <- rep(0:1, c(nv, n))
```

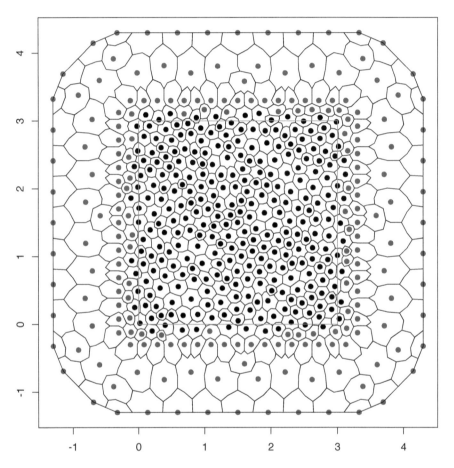

FIGURE 4.3 Voronoy polygons for the mesh used to make inference for the
log-Gaussian Cox process.

The exposure vector can be defined as:

```
e.pp <- c(w, rep(0, n))
```

The projection matrix is defined in two steps. For the integration points this is just a diagonal matrix because these locations are just the mesh vertices:

```
imat <- Diagonal(nv, rep(1, nv))
```

For the observed points, another projection matrix is defined:

```
lmat <- inla.spde.make.A(mesh, xy)
```

The entire projection matrix is:

```
A.pp <- rbind(imat, lmat)
```

We set up the data stack as follows:

```
stk.pp <- inla.stack(
  data = list(y = y.pp, e = e.pp),
  A = list(1, A.pp),
  effects = list(list(b0 = rep(1, nv + n)), list(i = 1:nv)),
  tag = 'pp')
```

Posterior marginals

The posterior marginals for all parameters of the model are obtained by fitting the model with INLA:

```
pp.res <- inla(y ~ 0 + b0 + f(i, model = spde),
  family = 'poisson', data = inla.stack.data(stk.pp),
  control.predictor = list(A = inla.stack.A(stk.pp)),
  E = inla.stack.data(stk.pp)$e)
```

The summary for the model hyperparameters, i.e., range and standard deviation of the spatial field, is:

```
pp.res$summary.hyperpar
##                  mean    sd 0.025quant 0.5quant 0.975quant   mode
```

```
## Range for i 2.225 1.472      0.636    1.836      6.093 1.307
## Stdev for i 0.332 0.107      0.161    0.319      0.579 0.294
```

The posterior marginal distributions of the log-Gaussian Cox model parameters are displayed in Figure 4.4.

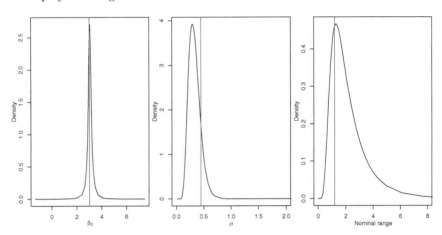

FIGURE 4.4 Posterior distribution for the parameters of the log-Gaussian Cox model β_0 (left), σ (center) and the nominal range (right). Vertical lines represent the actual values of the parameters.

4.2   Including a covariate in the log-Gaussian Cox process

In this section we add a covariate to the linear predictor in the log-Gaussian Cox model. To approximate the Cox likelihood well, we need the covariate to vary slowly in space; i.e. it should not change much between neighboring mesh nodes. For model fitting, we must know the value of the covariate at the location points and integration points.

4.2.1   Simulation of the covariate

First, we define a covariate everywhere in the study area, as

$$f(s_1, s_2) = \cos(s_1) - \sin(s_2 - 2).$$

This function will be computed at a grid defined from the settings in the spatstat package (e.g., the number of pixels in each direction). The locations where we simulate cover both the study window and the mesh points, as values of the covariate are required for both types of points:

```
# Use expanded range
x0 <- seq(min(mesh$loc[, 1]), max(mesh$loc[, 1]), length = npix)
y0 <- seq(min(mesh$loc[, 2]), max(mesh$loc[, 2]), length = npix)
gridcov <- outer(x0, y0, function(x,y) cos(x) - sin(y - 2))
```

Now, the expected number of points is a function of the covariate:

```
beta1 <- -0.5
sum(exp(beta0 + beta1 * gridcov) * diff(x0[1:2]) * diff(y0[1:2]))
## [1] 702.6
```

We simulate the point pattern using:

```
set.seed(1)
lg.s.c <- rLGCP('matern', im(beta0 + beta1 * gridcov, xcol = x0,
  yrow = y0), var = sigma2x, scale = range / sqrt(8),
  nu = 1, win = win)
```

Both the spatial field and the point pattern are returned. The point pattern locations are:

```
xy.c <- cbind(lg.s.c$x, lg.s.c$y)[, 2:1]
n.c <- nrow(xy.c)
```

The values of the simulated covariate and the simulated spatial field over the grid can be seen in Figure 4.5.

4.2.2 Inference

The covariate values need to be included in the data and the model for inference. These values must be available at the point pattern locations and at the mesh nodes. These values can be collected by interpolating from the grid data (in an im object) using function interp.im():

```
covariate.im <- im(gridcov, x0, y0)
covariate <- interp.im(covariate.im,
  x = c(mesh$loc[, 1], xy.c[, 1]),
  y = c(mesh$loc[, 2], xy.c[, 2]))
```

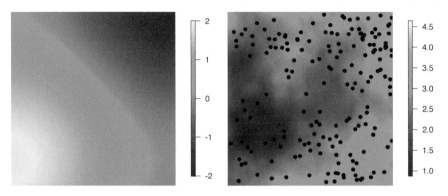

FIGURE 4.5 Simulated covariate (left) and simulated log-intensity of the point process, along with the simulated point pattern (right).

The augmented data is created in the same way as before:

```
y.pp.c <- rep(0:1, c(nv, n.c))
e.pp.c <- c(w, rep(0, n.c))
```

The projection matrix for the observed locations is:

```
lmat.c <- inla.spde.make.A(mesh, xy.c)
```

The two projection matrices can be merged with `rbind()` into a total projection matrix for the augmented data:

```
A.pp.c <- rbind(imat, lmat.c)
```

The data stack now includes the covariates, but is otherwise the same as in Section 4.1.3:

```
stk.pp.c <- inla.stack(
  data = list(y = y.pp.c, e = e.pp.c),
  A = list(1, A.pp.c),
  effects = list(list(b0 = 1, covariate = covariate),
    list(i = 1:nv)),
  tag = 'pp.c')
```

Finally, the model is fitted with the following R code:

```
pp.c.res <- inla(y ~ 0 + b0 + covariate + f(i, model = spde),
  family = 'poisson', data = inla.stack.data(stk.pp.c),
  control.predictor = list(A = inla.stack.A(stk.pp.c)),
  E = inla.stack.data(stk.pp.c)$e)
```

A summary of the model hyperparameters is given below.

```
pp.c.res$summary.hyperpar
##                 mean    sd 0.025quant 0.5quant 0.975quant   mode
## Range for i 2.196 1.38       0.651    1.846       5.804 1.344
## Stdev for i 0.404 0.14       0.188    0.385       0.733 0.347
```

The posterior distributions of the log-Gaussian Cox model parameters are shown in Figure 4.6.

4.3 Geostatistical inference under preferential sampling

In some cases, the sampling effort depends on the response (the marks or observed values at the points). For example, it is more common to have stations collecting data about pollution in industrial areas than in rural ones. This type of sampling is called *preferential sampling*. To make inference in this case, it is necessary to test whether there is a preferential sampling problem with our data. One approach is to build a joint model considering a log-Gaussian Cox model for the point pattern (the locations) and the response (Diggle et al., 2010). Hence, inference on a joint model is required in this case.

This approach assumes a linear predictor for the point process as:

$$\eta_i^{pp} = \beta_0^{pp} + u_i.$$

For the observations, the linear predictor is a function of the latent Gaussian random field for the point process:

$$\eta_i^y = \beta_0^y + \beta u_i$$

where β_0^y is an intercept for the observations and β is a weight on the shared random field effect.

An illustration of the use of INLA for the preferential sampling problem is available at http://www.r-inla.org/examples/case-studies/diggle09 in the R-INLA web page. The example therein uses a two dimensional random

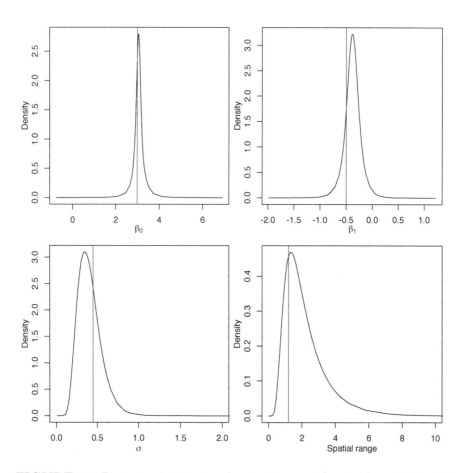

FIGURE 4.6 Posterior distribution for the intercept (top left), coefficient of the covariate (top right) and the parameters of the log-Cox model σ (bottom left) and nominal range (bottom right). Vertical lines represent the actual values of the parameters.

walk model for the latent random field. Here, the model considers geostatistical inference under preferential sampling using SPDE.

To simulate this joint likelihood we have to use the same underlying spatial field when simulating the point pattern and when simulating the observations. The y-values assigned to a point pattern location will use the value of the random field at the nearest grid point. The values of the spatial field are collected at the closest grid centers as follows:

```
z <- log(t(Lam$v)[do.call('cbind',
   nearest.pixel(xy[, 1], xy[, 2], Lam))])
```

We summarize these values:

```
summary(z)
##    Min.  1st Qu.  Median    Mean  3rd Qu.    Max.
##    2.09    2.87    3.14    3.20    3.62    4.37
```

These values are the latent field with zero mean plus the defined intercept β_0^{pp}. The response (the marks) is defined to have a different intercept β_0^y plus the zero mean random field multiplied by $1/\beta$, where β is the weight on the shared random field between the intensity of the point process locations and the response. Then, the response is simulated as follows:

```
beta0.y <- 10
beta <- -2
prec.y <- 16

set.seed(2)
resp <- beta0.y + (z - beta0) / beta +
   rnorm(length(z), 0, sqrt(1 / prec.y))
```

A summary of the simulated values of the response is below:

```
summary(resp)
##    Min.  1st Qu.  Median    Mean  3rd Qu.    Max.
##    8.85    9.63    9.94    9.90    10.22    10.69
```

As $\beta < 0$, the response values are lower where we have more observation locations. In a real data application this might be the case if there is a preference for survey locations where low values are expected.

4.3.1   Fitting the usual model

Here, a geostatistical model is fitted using the usual approach; i.e., the locations are considered as fixed. The mesh used to define the SPDE model is the same one as in the previous section. Hence, the data stack and the model can be fitted as follows:

```
stk.u <- inla.stack(
  data = list(y = resp),
  A = list(lmat, 1),
  effects = list(i = 1:nv, b0 = rep(1, length(resp))))

u.res <- inla(y ~ 0 + b0 + f(i, model = spde),
  data = inla.stack.data(stk.u),
  control.predictor = list(A = inla.stack.A(stk.u)))
```

A summary of the estimated values of the model parameters and the true values used to simulate the data are in Table 4.1. There, $1/\sigma_y^2$ denotes the precision of the Gaussian observations.

TABLE 4.1: Posterior modes of some of the model parameters.

Parameter	True	Mean	St. Dev.	2.5% quant.	97.5% quant.
β_0^y	10	9.946	0.1333	9.66	10.21
$1/\sigma_y^2$	16	14.476	1.7328	11.31	18.12

In addition, the posterior distribution marginals of the model parameters are in Figure 4.7.

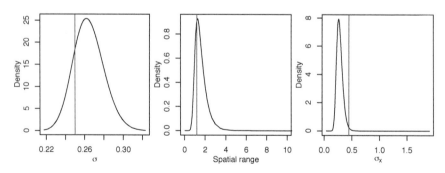

FIGURE 4.7 Posterior distribution for σ, the range and standard deviation of the spatial effect just using the response. Vertical lines represent the actual values of the parameters.

4.3.2 Model fitting under preferential sampling

Under preferential sampling, the fitted model is defined so that a LGRF is considered to model both point pattern and the response. Using INLA this can be done using two likelihoods: one for the point pattern and another one for the response. To do it, data must include a matrix response and a new index set to specify the model for the LGRF. This can be easily done by using the inla.stack() function following previous examples for models with two likelihoods.

The point pattern 'observations' are put in the first column and the response values in the second column. So, the data stack is redefined for the response and the point process. The response will be in the first column and the Poisson data for the point process in the second column. Also, to avoid the expected number of cases as NA for the Poisson likelihood, it is set as zero in the response part of the data stack. The SPDE effect in the point process part is modeled as a copy of the SPDE effect in the response part. This is done by defining an index set with a different name and using it in the copy feature later. See the code below.

```
stk2.y <- inla.stack(
  data = list(y = cbind(resp, NA), e = rep(0, n)),
  A = list(lmat, 1),
  effects = list(i = 1:nv, b0.y = rep(1, n)),
  tag = 'resp2')

stk2.pp <- inla.stack(data = list(y = cbind(NA, y.pp), e = e.pp),
  A = list(A.pp, 1),
  effects = list(j = 1:nv, b0.pp = rep(1, nv + n)),
  tag = 'pp2')

j.stk <- inla.stack(stk2.y, stk2.pp)
```

Now, the geostatistical model is fitted under preferential sampling. To include the LGRF in both likelihoods, the copy effect in INLA is used. We assign a $N(0, 2^{-1})$ prior for this parameter as follows:

```
# Gaussian prior
gaus.prior <- list(prior = 'gaussian', param = c(0, 2))
# Model formula
jform <- y ~ 0 + b0.pp + b0.y + f(i, model = spde) +
  f(j, copy = 'i', fixed = FALSE,
    hyper = list(theta = gaus.prior))
# Fit model
j.res <- inla(jform, family = c('gaussian', 'poisson'),
```

```
data = inla.stack.data(j.stk),
E = inla.stack.data(j.stk)$e,
control.predictor = list(A = inla.stack.A(j.stk)))
```

Values of the parameters used in the simulation of the data and a summary of
their posterior distributions are shown in Table 4.2.

TABLE 4.2: Posterior modes of some of the model parameters under
preferential sampling.

Parameter	True	Mean	St. Dev.	2.5% quant.	97.5% quant.
β_0	3	3.048	0.1955	2.659	3.453
β_0^y	10	9.961	0.1485	9.645	10.254

Posterior marginal distributions for the model parameters from the result
considering only the point process (PP), only the observations/marks ($\mathbf{Y}$) and
jointly are in Figure 4.8. Notice that for the β_0 parameter there are results
considering only the PP and joint, for β_0^y the results are only for $\mathbf{Y}$ and the
joint model, and for β (fitted using copy) the results are only from the joint
model. Results from the three models are only available for the random field
parameters.

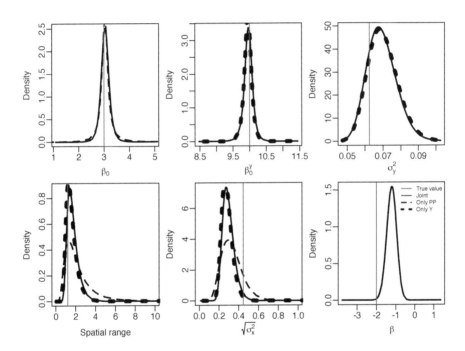

FIGURE 4.8 Posterior marginal distribution for the intercept for the point process β_0, intercept for the observations β_0^y, noise variance in the observations σ_y^2, the practical range, the marginal standard deviation for the random field σ_x^2 and the sharing coefficient β.

5

Spatial non-stationarity

In this chapter we focus on non-stationary models. In particular, these models will show how to model the covariance using covariates and how to deal with physical barriers in the study region.

5.1 Explanatory variables in the covariance

In this section we present an example of the model proposed in Ingebrigtsen et al. (2014). This example describes a way to include explanatory variables (i.e., covariates) in both of the SPDE model parameters. We will consider the parametrization used along this book and detailed in Lindgren et al. (2011).

5.1.1 Introduction

First of all, we remind ourselves of the definition for the precision matrix. Restricting to $\alpha = 1$ and $\alpha = 2$, the precision matrix is given as follows:

- $\alpha = 1$: $\mathbf{Q}_{1,\tau,\kappa} = \tau(\kappa^2 \mathbf{C} + \mathbf{G})$
- $\alpha = 2$: $\mathbf{Q}_{2,\tau,\kappa} = \tau(\kappa^4 \mathbf{C} + \kappa^2 \mathbf{G} + \kappa^2 \mathbf{G} + \mathbf{G}\mathbf{C}^{-1}\mathbf{G})$

The default stationary SPDE model implemented in `inla.spde2.matern()` function considers the θ_1 as the log of the local precision parameter τ and θ_2 the log of the scale κ parameter. The marginal variance σ^2 and the range ρ are function of these parameters as

$\sigma^2 = \Gamma(\nu)/(\Gamma(\alpha)(4\pi)^{d/2}\kappa^{2\nu}\tau^2)$ and $\rho = \sqrt{8\nu}/\kappa$. These relations give

$$\begin{aligned} \log(\tau) &= \tfrac{1}{2}\log\left(\frac{\Gamma(\nu)}{\Gamma(\alpha)(4\pi)^{d/2}}\right) - \log(\sigma) - \nu\log(\kappa) \\ \log(\kappa) &= \frac{\log(8\nu)}{2} - \log(\rho). \end{aligned} \tag{5.1}$$

The approach to non-stationarity proposed in Ingebrigtsen et al. (2014) is to consider a regression like model for $\log\tau$ and $\log\kappa$. It is done considering basis functions as

$$
\begin{aligned}
\log(\tau(\mathbf{s})) &= b_0^{(\tau)}(\mathbf{s}) + \sum_{k=1}^{p} b_k^{(\tau)}(\mathbf{s})\theta_k \\
\log(\kappa(\mathbf{s})) &= b_0^{(\kappa)}(\mathbf{s}) + \sum_{k=1}^{p} b_k^{(\kappa)}(\mathbf{s})\theta_k
\end{aligned}
\tag{5.2}
$$

where we have the θ_k parameters as regression on the basis functions. Setting $\mathbf{T}$ as a diagonal matrix with elements as $\tau(\mathbf{s})$ and $\mathbf{K}$ a diagonal matrix with elements as $\kappa(\mathbf{s})$ we now have the precision matrix as

- $\alpha = 1$: $\mathbf{Q}_{1,\theta,\mathbf{B}^{(\tau)},\mathbf{B}^{(\kappa)}} = \mathbf{T}(\mathbf{K}\mathbf{C}\mathbf{K} + \mathbf{G})\mathbf{T}$

- $\alpha = 2$: $\mathbf{Q}_{2,\theta,\mathbf{B}^{(\tau)},\mathbf{B}^{(\kappa)}} = \mathbf{T}(\mathbf{K}^2\mathbf{C}\mathbf{K}^2 + \mathbf{K}\mathbf{G} + \mathbf{G}^T\mathbf{K} + \mathbf{G}\mathbf{C}^{-1}\mathbf{G})\mathbf{T}$

These basis functions are passed in the argument `B.tau` and `B.kappa` arguments of the `inla.spde2.matern()` function. $b_0^{(\tau)}$ goes in the first column of `B.tau` and $b_k^{(\tau)}$ in the next columns. $b_0^{(\kappa)}$ goes in the first column of `B.kappa` and $b_k^{(\tau)}$ in the next columns. When these basis matrices are supplied as just one line matrix, the actual basis matrix will be formed having all lines equal to this unique line matrix and the model is stationary. The default uses $[0\ 1\ 0]$ (one by three) matrix for the local precision parameter τ and $[0\ 0\ 1]$ (one by three) matrix for the scaling parameter κ. Thus θ_1 controls $\log(\tau)$ and θ_2 controls $\log(\kappa)$.

We can use `B.tau` and `B.kappa` in such a way to build a model for the marginal standard deviation and range, see Lindgren and Rue (2015). This is the case of the function `inla.spde2.pcmatern()`, where `B.tau` is $[\log(\tau_0)\ \nu\ -1]$ and `B.kappa` is $[\log(\kappa_0)\ -1\ 0]$. We are using this along this book having the range as the first parameter and σ the second.

Both the marginal standard deviation and the range can be modeled considering a regression model as detailed in Lindgren and Rue (2015). This is the case of considering that the log of σ and log of ρ are modeled by a regression on basis functions as

$$
\log(\sigma(\mathbf{s})) = b_0^{\sigma}(\mathbf{s}) + \sum_{k=1}^{p} b_k^{\sigma}(\mathbf{s})\theta_k
\tag{5.3}
$$

$$
\log(\rho(\mathbf{s})) = b_0^{\rho}(\mathbf{s}) + \sum_{k=1}^{p} b_k^{\rho}(\mathbf{s})\theta_k
\tag{5.4}
$$

where $b_0^{\sigma}(\mathbf{s})$ and $b_0^{\rho(\mathbf{s})}$ are offsets, $b_k^{\sigma}()$ and $b_k^{\rho()}$ are basis functions that can be defined on spatial locations or covariates, each one with an associated θ_k parameter. Notice that `B.tau` and `B.kappa` are basis functions evaluated at each mesh node. Therefore, to include a covariate in σ^2 or in the range we do need to have it available at each mesh node.

In our example we consider the domain area as the rectangle $(0, 10) \times (0, 5)$

and the range as a function of the first coordinate of the location defined by $\rho(\mathbf{s}) = \exp(\theta_2 + \theta_3(s_{i,1} - 5)/10)$. This gives

$$\begin{aligned} \log(\sigma) &= \log(\sigma_0) + \theta_1 \\ \log(\rho(\mathbf{s})) &= \log(\rho_0) + \theta_2 + \theta_3 b(\mathbf{s}) \end{aligned} \quad (5.5)$$

where $b(\mathbf{s}) = (s_{i,1} - 5)/10$.

We now define the values for θ_1, θ_2 and θ_3. Because we choose θ_3 positive it implies an increasing range along the first location coordinate. The range as function of the first coordinate is shown in Figure 5.1.

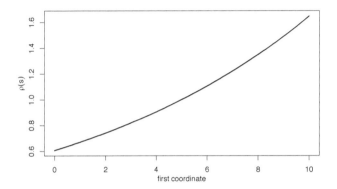

FIGURE 5.1 Range as function of the first coordinate.

We have to define the basis functions for $\log(\tau)$ and $\log(\kappa)$. Following Lindgren and Rue (2015) we need to write $\log(\kappa(\mathbf{s}))$ and $\log(\tau(\mathbf{s}))$ as functions of $\log(\rho(\mathbf{s}))$ and $\log(\sigma)$. Substituting $\log(\sigma)$ and $\log(\rho(\mathbf{s}))$ into the equations for $\log(\kappa(\mathbf{s}))$ and $\log(\tau(\mathbf{s}))$ we have

$$\begin{aligned} \log(\tau(\mathbf{s})) &= \log(\tau_0) - \theta_1 + \nu\theta_2 + \nu\theta_3 b(\mathbf{s}) \\ \log(\kappa(\mathbf{s})) &= \log(\kappa_0) - \theta_2 - \theta_3 b(\mathbf{s}) \ . \end{aligned} \quad (5.6)$$

Therefore we have to include $(s_{i,1} - 5)/10$ as a fourth column in B.tau and minus it in B.kappa as well.

5.1.2 Implementation of the model

First, we define the $(0, 10) \times (0, 5)$ rectangle:

```
pl01 <- cbind(c(0, 1, 1, 0, 0) * 10, c(0, 0, 1, 1, 0) * 5)
```

The mesh will be created using this polygon, as follows:

```
mesh <- inla.mesh.2d(loc.domain = pl01, cutoff = 0.1,
  max.edge = c(0.3, 1), offset = c(0.5, 1.5))
```

This mesh is made of 2196 nodes.

We supply $(s_{i,1} - 5)/10$ as a fourth column of in the B.tau argument of the inla.spde2.matern() function, and minus it in the fourth column of argument B.kappa. In addition, it is also necessary to set a prior distribution on the vector of hyperparameters θ according to its new dimension, a three length vector. The default is a Gaussian distribution, for which mean and precision diagonal need to be specified in two vectors, as follows:

```
nu <- 1
alpha <- nu + 2 / 2
# log(kappa)
logkappa0 <- log(8 * nu) / 2
# log(tau); in two lines to keep code width within range
logtau0 <- (lgamma(nu) - lgamma(alpha) -1 * log(4 * pi)) / 2
logtau0 <- logtau0 - logkappa0
# SPDE model
spde <- inla.spde2.matern(mesh,
  B.tau = cbind(logtau0, -1, nu, nu * (mesh$loc[,1] - 5) / 10),
  B.kappa = cbind(logkappa0, 0, -1, -1 * (mesh$loc[,1] - 5) / 10),
  theta.prior.mean = rep(0, 3),
  theta.prior.prec = rep(1, 3))
```

The precision matrices are built with

```
Q <- inla.spde2.precision(spde, theta = theta)
```

5.1.3   Simulation at the mesh nodes

We can draw one realization of the process using the inla.qsample() function as

```
sample <- as.vector(inla.qsample(1, Q, seed = 1))
```

The top plot in Figure 5.2 shows the simulated values projected to a grid Here, the projector matrix has been used to project the simulated values in the grid limited in the unit square with limits (0, 10) and (0, 5). The bottom plot in Figure 5.2 shows the spatial correlation with respect to two points, computed using function book.spatial.correlation(). The interesting issue to observe here is how the range at which spatial autocorrelation disappears (i.e., correlation is zero) is smaller around the point on the left due to the increasing range with the x-coordinate simulated in the model.

The results and plots in Figure 5.2 have been computed with this code:

```
# Plot parameters
par(mfrow = c(2, 1), mar = c(0, 0, 0, 0))
#Plot Field
proj <- inla.mesh.projector(mesh, xlim = 0:1 * 10, ylim = 0:1 * 5,
  dims = c(200, 100))
book.plot.field(sample, projector = proj)
# Compute spatial autocorrelation
cx1y2.5 <- book.spatial.correlation(Q, c(1, 2.5), mesh)
cx7y2.5 <- book.spatial.correlation(Q, c(7, 2.5), mesh)
# Plot spatial autocorrelation
book.plot.field(cx1y2.5, projector = proj, zlim = c(0.1, 1))
book.plot.field(cx7y2.5, projector = proj, zlim = c(0.1, 1),
  add = TRUE)
```

5.1.4   Estimation with data simulated at the mesh nodes

A model can be easily fitted with the data simulated at mesh nodes. Considering that there are observations exactly at each mesh node, there is no need to use any predictor matrix and the stack functionality. Because data are just realizations from the random field, there is no noise and then the precision of the Gaussian likelihood will be fixed to a very large value. For example, to value $\exp(20)$:

```
clik <- list(hyper = list(theta = list(initial = 20,
  fixed = TRUE)))
```

Because the random field has a zero mean, there are not fixed parameters to fit. Hence, model fitting can be done as follows:

```
formula <- y ~ 0 + f(i, model = spde)

res1 <- inla(formula, control.family = clik,
  data = data.frame(y = sample, i = 1:mesh$n))
```

Now, the summary of the posterior marginal distribution for θ (joined with the true values) is provided in Table 5.1. The results obtained are quite good and the point estimates are very close to the actual values.

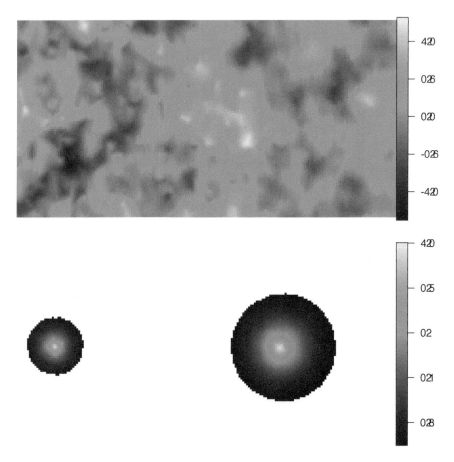

FIGURE 5.2 The simulated random field with increasing range along the horizontal coordinate (top) and the correlation at two location points: $(1, 2.5)$ and $(7, 2.5)$ (bottom).

TABLE 5.1: Summary of the posterior distributions of the parameters in the non-stationary example for the data simulated at the mesh nodes.

Parameter	True	Mean	St. Dev.	2.5% quant.	97.5% quant.
θ_1	-1	-0.9745	0.0339	-1.0409	-0.9072
θ_2	0	0.0102	0.0435	-0.0752	0.0960
θ_3	1	1.0937	0.0607	0.9813	1.2196

5.1.5 Estimation with locations not at the mesh nodes

The more general case is when observations are not at the mesh nodes. For this reason, some in the next example data will be at the locations simulated with the following R code:

```
set.seed(2)
n <- 200
loc <- cbind(runif(n) * 10, runif(n) * 5)
```

Next, we take the simulated random field at the mesh vertices and project it into these locations. For this, a projector matrix is needed:

```
projloc <- inla.mesh.projector(mesh, loc)
```

Hence, the projection is:

```
x <- inla.mesh.project(projloc, sample)
```

This provides the sample data at the simulated locations.

Now, because these locations are not vertices of the mesh, the stack functionality is required to put all the data together. First, the predictor matrix is needed, but this is the same used to sample the data.

Then, a stack is defined for this sample:

```
stk <- inla.stack(
  data = list(y = x),
  A = list(projloc$proj$A),
  effects = list(data.frame(i = 1:mesh$n)),
  tag = 'd')
```

Finally, the model is fitted:

```
res2 <- inla(formula, data = inla.stack.data(stk),
  control.family = clik,
  control.predictor = list(compute = TRUE, A = inla.stack.A(stk)))
```

The true values and a summary of marginal posterior distributions for θ are given in Table 5.2. As in the previous example, estimates are quite close to the actual values of the parameters used in the simulation.

TABLE 5.2: Summary of the posterior distributions of the parameters in the non-stationary example for the data simulated at the locations.

Parameter	True	Mean	St. Dev.	2.5% quant.	97.5% quant.
θ_1	-1	-0.9440	0.0653	-1.0702	-0.8134
θ_2	0	-0.1906	0.1204	-0.4219	0.0517
θ_3	1	0.9571	0.3182	0.3293	1.5810

Figure 5.3 shows the simulated values, the predicted (posterior mean) and the projected posterior standard deviation. Here, it can be seen how the predicted values are similar to the simulated true values.

5.2 The Barrier model

The most common spatial models are the stationary isotropic ones. For these, moving or rotating the map does not change the model. Even though this is a reasonable assumption in many cases, it is not a reasonable assumption if there are physical barriers in the study area. For example, when we model aquatic species near the coast, a stationary model cannot be aware of the coastline, and a new model is needed (one which is aware of the coastline).

In the paper by Bakka et al. (2016) a new non-stationary model was constructed for use in INLA, with a syntax very similar to the stationary model (to make it easy to use). In this section we give an example using this model on a part of the Canadian coastline.

5.2.1 Canadian coastline example

The example developed here is based on simulated data using the Canadian coastline as a physical barrier to spatial correlation. This example requires

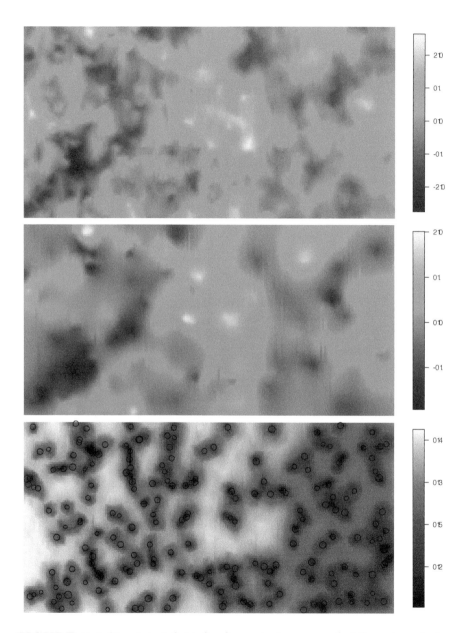

FIGURE 5.3 Simulated field (top), posterior mean (mid) and posterior standard deviation with the location points added (bottom).

some manipulation of maps and polygons in order to create an appropriate mesh for the model, and we have loaded additional packages for this.

Polygon

First, we construct some spatial polygon covering our study area. Here we do this via the `mapdata` package (Becker et al., 2016):

```
# Select region
map <- map("world", "Canada", fill = TRUE,
  col = "transparent", plot = FALSE)
IDs <- sapply(strsplit(map$names, ":"), function(x) x[1])
map.sp <- map2SpatialPolygons(
  map, IDs = IDs,
  proj4string = CRS("+proj=longlat +datum=WGS84"))
```

Next, we define our study area `pl.sel`, as a manually constructed polygon, and intersect this with the coastal area. Since we have a polygon for land, we take the difference instead of an intersection, as follows:

```
pl.sel <- SpatialPolygons(list(Polygons(list(Polygon(
  cbind(c(-69, -62.2, -57, -57, -69, -69),
    c(47.8, 45.2, 49.2, 52, 52, 48)),
    FALSE)), '0')), proj4string = CRS(proj4string(map.sp)))

poly.water <- gDifference(pl.sel, map.sp)
```

The Canadian coastline and the manually constructed polygon can be seen on the left plot in Figure 5.4.

Transforming to UTM coordinates

So far we have used longitude and latitude in this example, but this is not a valid coordinate system for spatial modeling, as distances in longitude and distances in latitude are in different scales. Modeling should be done in a CRS (Coordinate Reference System) where units along each axis is measured in e.g. kilometers. We do not want to use meters as this would result in very large values along the axes, and could cause unstable numerical results.

The projection is done with function `spTransform()` as follows:

```
# Define UTM projection
kmproj <- CRS("+proj=utm +zone=20 ellps=WGS84 +units=km")
# Project data
poly.water = spTransform(poly.water, kmproj)
```

```
pl.sel = spTransform(pl.sel, kmproj)
map.sp = spTransform(map.sp, kmproj)
```

Simple mesh

Before we construct the mesh we are going to use, we first show how to construct a mesh only in water. We then discuss the difference between the two approaches.

```
mesh.not <- inla.mesh.2d(boundary = poly.water, max.edge = 30,
  cutoff = 2)
```

The created mesh has 1106 nodes. This mesh has been plotted in Figure 5.4 (right plot).

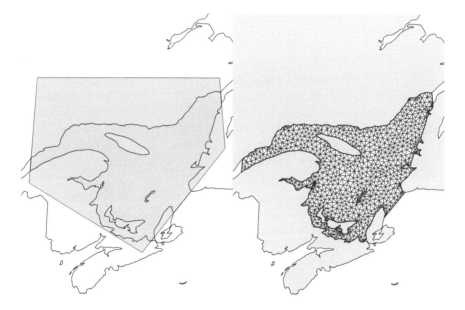

FIGURE 5.4 The left plot shows the polygon for land in grey and the manually constructed polygon for our study area in light blue. The right plot shows the simple mesh, constructed only in the water.

Before developing the Barrier model, we used to construct this simple mesh and use the SPDE model here, as this avoids smoothing across land. The problem with this approach, however, is that we introduce other hidden assumptions, called Neumann boundary conditions. As shown by Bakka et al. (2016), this assumption can lead to even worse behavior than the stationary model (which smooths over islands).

Mesh over land and sea

Next, we construct a mesh over both water and land. This is the mesh we are going to use. We include the coastline polygon in the function call, to make the edges of the triangles follow the coastline.

```
max.edge = 30
bound.outer = 150
mesh <- inla.mesh.2d(boundary = poly.water,
  max.edge = c(1,5) * max.edge,
  cutoff = 2,
  offset = c(max.edge, bound.outer))
```

Next, we select the triangles of the mesh that are inside `poly.water`. We define the triangles in the barrier area (i.e. on land) to be all the triangles that are not on water. From this we construct the `poly.barrier`. This should match the original land polygon closely, but will deviate a little. The new polygon `poly.barrier` is where our model assumes there to be land; hence we use this polygon also for plotting the results.

```
water.tri = inla.over_sp_mesh(poly.water, y = mesh,
  type = "centroid", ignore.CRS = TRUE)
num.tri = length(mesh$graph$tv[, 1])
barrier.tri = setdiff(1:num.tri, water.tri)
poly.barrier = inla.barrier.polygon(mesh,
  barrier.triangles = barrier.tri)
```

The mesh in Figure 5.5 is different from the simple mesh in Figure 5.4, as we have also defined the mesh over land. We can use this mesh to construct both a stationary model and a Barrier model.

The Barrier model vs. the stationary model

Next we define the precision matrix for the Barrier model and the stationary model

$$\mathbf{u} \sim \mathcal{N}(0, Q^{-1})$$

considering unit marginal variance:

```
range <- 200
barrier.model <- inla.barrier.pcmatern(mesh,
  barrier.triangles = barrier.tri)
Q <- inla.rgeneric.q(barrier.model, "Q", theta = c(0, log(range)))
```

Constrained refined Delaunay triangulation

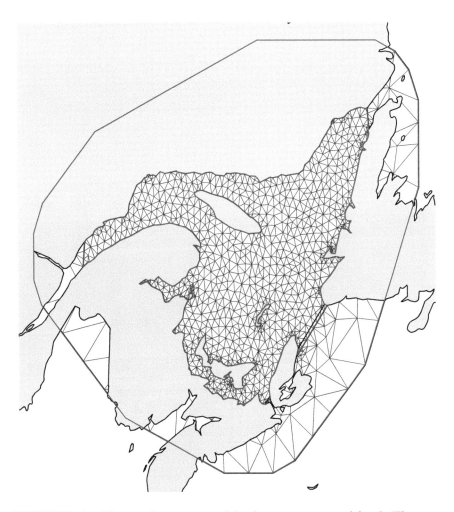

FIGURE 5.5 The mesh constructed both over water and land. The grey region is the original land map. The inner red outline marks the coastline barrier.

This code spits out a warning because the priors are not defined. However, as we are simulating from the GF, the priors for the hyperparameters are irrelevant.

```
stationary.model <- inla.spde2.pcmatern(mesh,
  prior.range = c(1, 0.1), prior.sigma = c(1, 0.1))
Q.stat <- inla.spde2.precision(stationary.model,
  theta = c(log(range), 0))
```

The priors we chose here are arbitrary as they are not used. Note that the ordering of `theta` is reversed between the two models.

We use the functions from `spde-book-functions.R` to compute spatial correlation surfaces; see Figure 5.6.

```
# The location we find the correlation with respect to
loc.corr <- c(500, 5420)
corr <- book.spatial.correlation(Q, loc = loc.corr, mesh)
corr.stat <- book.spatial.correlation(Q.stat, loc = loc.corr,
  mesh)
```

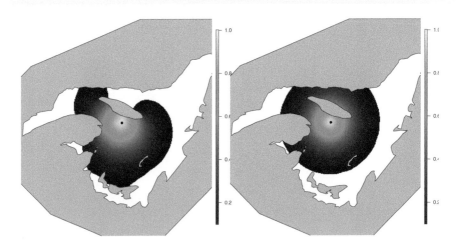

FIGURE 5.6 The left plot shows the correlation structure of the Barrier model, with respect to the black point, while the right plot shows the correlation structure of the stationary model.

Sample from the Barrier model

Next, we sample locations from the water polygon using `spsample()`. And we simplify the data structure `loc.data` from `SpatialPoints` to just a matrix of values.

```
set.seed(201805)
loc.data <- spsample(poly.water, n = 1000, type = "random")
loc.data <- loc.data@coords
```

We sample the continuous field **u** and project it to data locations:

```
# Seed is the month the code was first written times some number
u <- inla.qsample(n = 1, Q = Q, seed = 201805 * 3)[, 1]
A.data <- inla.spde.make.A(mesh, loc.data)
u.data <- A.data %*% u

# df is the dataframe used for modeling
df <- data.frame(loc.data)
names(df) <- c('locx', 'locy')
# Size of the spatial signal
sigma.u <- 1
# Size of the measurement noise
sigma.epsilon <- 0.1
df$y <- drop(sigma.u * u.data + sigma.epsilon * rnorm(nrow(df)))
```

Inference with the Barrier model

Similarly as with previous models, we construct the typical stack object to prepare the data for model fitting:

```
stk <- inla.stack(
  data = list(y = df$y),
  A = list(A.data, 1),
  effects =list(s = 1:mesh$n, intercept = rep(1, nrow(df))),
  tag = 'est')
```

The formula for the model only takes an intercept plus the spatial effect:

```
form.barrier <- y ~ 0 + intercept + f(s, model = barrier.model)
```

Finally, we run INLA with a Gaussian likelihood:

```
res.barrier <- inla(form.barrier, data = inla.stack.data(stk),
  control.predictor = list(A = inla.stack.A(stk)),
  family = 'gaussian',
  control.inla = list(int.strategy = "eb"))
```

Posterior spatial field

In Figure 5.7 we compare the results from the Barrier model to the true (simulated) spatial field. These have a close match. We observe quick changes across islands or small peninsulas, which may be smoothed out by a stationary model. See Bakka et al. (2016) for a more detailed discussion.

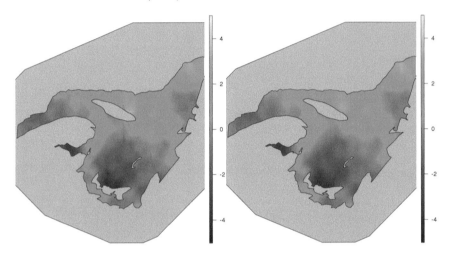

FIGURE 5.7 The left plot shows the true simulated spatial field **u**, while the right plot shows the posterior mean of the Barrier model.

Posterior of hyperparameters

The summary of the posterior marginals of the model hyperparameters is:

```
res.barrier$summary.hyperpar
##                                          mean      sd
## Precision for the Gaussian observations 95.016 6.2467
## Theta1 for s                            0.169 0.1583
## Theta2 for s                            5.516 0.1723
##                                          0.025quant 0.5quant
## Precision for the Gaussian observations   83.2403  94.8421
## Theta1 for s                              -0.1059   0.1552
## Theta2 for s                               5.2172   5.5010
##                                          0.975quant    mode
## Precision for the Gaussian observations   107.8314  94.538
## Theta1 for s                                0.5109   0.104
## Theta2 for s                                5.8884   5.445
```

To summarize or plot the hyperparameters in the Barrier model, we must know the order they come in and the transformations:

$$\sigma = e^{\theta_1} \quad \text{is the marginal standard deviation} \quad (5.7)$$

$$r = e^{\theta_2} \quad \text{is the spatial range} \quad (5.8)$$

From this we compute Table 5.3. We see that our estimates and intervals recover the true values.

TABLE 5.3: Summary of the true values and the posterior of the hyperparameters in the Barrier model.

Parameter	True	50% quant.	2.5% quant.	97.5% quant.
σ	1	1.168	0.8995	1.667
range	200	244.941	184.4189	360.811

5.3 Barrier model for noise data in Albacete (Spain)

Barrier models can be appropriate tools to model different types of environmental phenomena that have a clear anisotropic behavior. A clear example is the propagation of noise in cities, where buildings act as barriers in the propagation of noise. In this section, spatial and spatio-temporal models are developed using noise data from the city of Albacete (Castilla-La Mancha, Spain).

Data has been extracted from a report commissioned by the local authorities due to increasing concern about noise in the city and, in particular, in certain areas of the city center. The analysis will focus on a busy area in the city center (popularly know as "La zona", i.e., "The zone") with plenty of bars and restaurants. The original reports are available (as PDF documents) from website `http://www.pioneraconsultores.com/es/mapa-de-ruidos.zhtm?corp=medioambiente`, which contains the original data.

Measurements were taken in March 2010 for a period of 24 hours at different points throughout the city. In this analysis, hourly measurements of sound pressure levels (in A-weighted decibels, or dbA) at 7 different points within the study region will be considered. Local regulations require noise levels to be below 65 db during the day and below 60 db at night in residential areas. Hence, the aim of this example is to study how noise levels vary within the city center to get an idea of whether local regulations are fulfilled.

The data used in this example is provided as a `SpatialPointsDataFrame` in file `noise.RData`. Table 5.4 summarizes the main variables in the dataset.

TABLE 5.4: Variables in the dataset on noise measurements in Albacete (Spain).

Variable	Description
X	x-coordinate (in UTM).
Y	y-coordinate (in UTM).
LAeqZZh	Hourly sound pressure level (in dbA) between ZZ and ZZ + 1 hours. ZZ is an integer from 1 to 24.

Data loading

Data about the locations of the buildings, which will act as barriers, will be obtained from OpenStreetMap (OSM, OpenStreetMap contributors, 2018) using package osmar (Eugster and Schlesinger, 2013). This package provides a simple way to download OSM data and extract the relevant information required to build the desired models.

In the next lines of R code, function osmsource_api() sets up access to the OSM API. Next, function center_bbox is used to set the squared region where the data is extracted from. The first two arguments are the coordinates of the center of the rectangle, and the next two its width and height. Finally, function get_osm() gets the data from the defined bounding box using the OSM API.

```
#Get building data from Albacete using OpenStreetMap
library(osmar)

src <- osmsource_api()

bb <- center_bbox(-1.853152, 38.993318,  400, 400)
ua <- get_osm(bb, source = src)
```

OSM data provides a wealth of information about street data in variable ua. Since our focus is on modeling noise using buildings as barriers, the information about buildings will be extracted. Functions find() and find_down() are used below to obtain an index of all features in the OSM data in ua tagged as building. Then subset() is used to select the elements in ua that are in the index of buildings. Finally, function as_sp converts the boundaries of the buildings (as associated information) into a SpatialPolygonsDataFrame.

```
idx <- find(ua, way(tags(k == "building")))
idx <- find_down(ua, way(idx))
bg <- subset(ua, ids = idx)
```

```
bg_poly <- as_sp(bg, "polygons")
```

The newly created `SpatialPolygonsDataFrame` object stores different buildings as separate entities. Hence, we will use function `unionSpatialPolygons()` to create a single entity, which will be used later to obtain the boundaries of the streets.

```
library(maptools)
bg_poly <- unionSpatialPolygons(bg_poly,
  rep("1", length(bg_poly)))
```

Next, the actual study area is defined by using function `center_bbox()` again, this time with a smaller width and height, to make sure that all buildings in the study area have been retrieved before. Finally, a `SpatialPolygons` object is created and the intersection between the boundary of the study area and the polygons of the buildings is created using function `gIntersection` from the `rgeos` package (Bivand and Rundel, 2017).

```
#Outer boundary to overlay with buildings
bb.outer <- center_bbox(-1.853152, 38.993318,  350, 350)
pl <- matrix(c(bb.outer[1:2], bb.outer[c(3, 2)], bb.outer[3:4],
  bb.outer[c(1, 4)]), ncol = 2, byrow = TRUE)

pl_sp <- SpatialPolygons(
  list(Polygons(list(Polygon(pl)), ID = 1)),
  proj4string = CRS(proj4string(bg_poly)))

library(rgeos)

bg_poly2 <- gIntersection(bg_poly, pl_sp, byid = TRUE)
```

Coordinates of the buildings are in longitude and latitude, which is not an adequate coordinate reference system to deal with distances. For this reason, coordinates will be converted into UTM, so that distances are measured in meters. For this, function `spTransform()` will be used:

```
## Transform data
library(rgdal)
pl_sp.utm30 <- spTransform(pl_sp,
  CRS("+proj=utm +zone=30 +ellps=GRS80 +units=m +no_defs"))
bg_poly2.utm30 <- spTransform(bg_poly2,
  CRS("+proj=utm +zone=30 +ellps=GRS80 +units=m +no_defs"))
```

The last part of the building data processing involves obtaining not the polygons of the buildings, but the polygons of the streets, which is where the spatial process takes place. This will be done by taking the bounding box of the study region and removing the buildings, as follows:

```
## Compute difference to get streets
bg_poly2.utm30 <- gDifference(pl_sp.utm30, bg_poly2.utm30)
```

Figure 5.8 shows the buildings (left plot) and the streets (right plot) that will be used in the analysis.

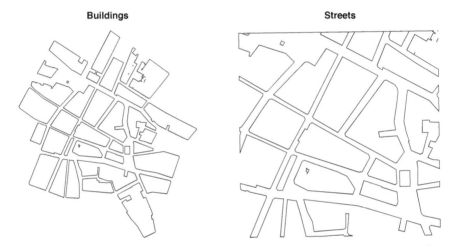

FIGURE 5.8 Building and street boundaries obtained from OpenStreetMap.

Noise data is provided in a `data.frame` in file `noise.RData`, and it can be loaded as:

```
load("data/noise.RData")
```

This dataset contains data from 7 locations where noise has been measured. This includes 5 outdoors locations plus 2 indoors locations. Although noise may be attenuated at these 2 locations they have been kept in the dataset to have more observations in the analysis.

Mesh building

The mesh will be built similarly to that in Section 5.2.1. In this case, distances are measured in meters. The maximum edges of the triangles will be 5 meters and the outer boundary will be 10 meters:

```
max.edge = 5
bound.outer = 10
mesh <- inla.mesh.2d(boundary = pl_sp.utm30,
  max.edge = c(1, 1.5) * max.edge,
  cutoff = 10,
  offset = c(max.edge, bound.outer))
```

The next lines of R code will use function `inla.over_sp_mesh()` (to check which mesh triangles are inside a polygon) and call function `inla.barrier.polygon()` to obtain the polygon around the barrier. Figure 5.9 shows the mesh created, together with the streets and the polygon around the barrier obtained with `inla.barrier.polygon()` (in red).

```
city.tri = inla.over_sp_mesh(bg_poly2.utm30, y = mesh,
  type = "centroid", ignore.CRS = TRUE)
num.tri <- length(mesh$graph$tv[, 1])
barrier.tri <- setdiff(1:num.tri, city.tri)
poly.barrier <- inla.barrier.polygon(mesh,
  barrier.triangles = barrier.tri)
```

Barrier model for noise data

Before fitting the actual barrier model to the noise data, we will explore how the spatial correlation in the barrier model is defined. First of all, a model with range 100 and precision 1 is simulated. Here, function `inla.barrier.pcmatern()` is used to define a barrier model using the `rgeneric` approach in INLA to define latent models; see `inla.doc("rgeneric")` for details. Furthermore, `inla.rgeneric.q()` returns the precision matrix of the model for some values of the hyperparameters, i.e., precision and range (in the log scale).

```
range <- 100
prec <- 1
barrier.model <- inla.barrier.pcmatern(mesh,
  barrier.triangles = barrier.tri)
Q <- inla.rgeneric.q(barrier.model, "Q",
  theta = c(log(prec), log(range)))
```

In order to compare the Barrier model, a stationary model is defined below. Function `inla.spde2.pcmatern()` defines a stationary model using a PC prior for the standard deviation and range, and function `inla.spde2.precision()` returns the precision matrix of the model.

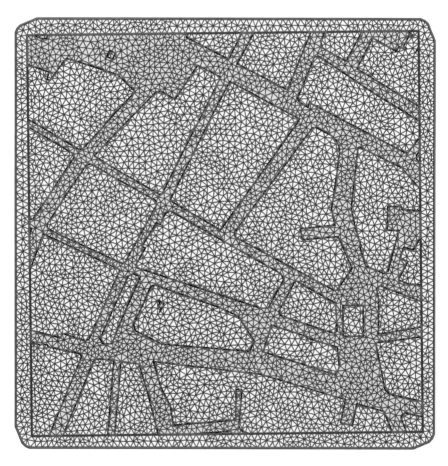

FIGURE 5.9 Mesh created for the analysis of noise data in Albacete (Spain).

```
stationary.model <- inla.spde2.pcmatern(mesh,
  prior.range = c(100, 0.9), prior.sigma = c(1, 0.1))
Q.stat <- inla.spde2.precision(stationary.model,
  theta = c(log(range), 0))
```

Then, the correlation field at point (599318.3, 4316661) is computed for the Barrier and stationary models:

```
# The location we find the correlation with respect to
loc.corr <- c(599318.3, 4316661)

corr <- book.spatial.correlation(Q, loc = loc.corr, mesh)
corr.stat = book.spatial.correlation(Q.stat, loc = loc.corr, mesh)
```

These are shown in Figure 5.10. Note how the spatial correlation for the stationary model (right plot in Figure 5.10) seems to ignore buildings and only considers Euclidean distance between two points, while the spatial autocorrelation for the Barrier model (left plot in Figure 5.10) does take buildings into account.

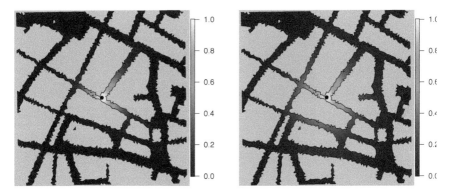

FIGURE 5.10 The left plot shows the correlation structure of the Barrier model, with respect to the black point, while the right plot shows the correlation structure of the stationary model.

The first Barrier model will consider the spatial variation of noise at 1 am, which is measured by variable **LAeq1h**. The projector matrix and stack required to fit the model can be obtained as follows:

```
A.data <- inla.spde.make.A(mesh, coordinates(noise))

stk <- inla.stack(
```

```
  data = list(y = noise$LAeq1h),
  A = list(A.data, 1),
  effects =list(s = 1:mesh$n, intercept = rep(1, nrow(noise))),
  tag = 'est')
```

Similarly, a new stack will be created for prediction at the mesh nodes:

```
# Projector matrix at prediction points
A.pred <- inla.spde.make.A(mesh, mesh$loc[, 1:2])

#Stack for prediction at mesh nodes
stk.pred <- inla.stack(
  data = list(y = NA),
  A = list(A.pred, 1),
  effects =list(s = 1:mesh$n, intercept = rep(1, nrow(A.pred))),
  tag = 'pred')

# Joint stack for model fitting and prediction
joint.stk <- inla.stack(stk, stk.pred)
```

Next, the formula to fit the model is:

```
form.barrier <- y ~ 0 + intercept + f(s, model = barrier.model)
```

Then, the model is fitted using a PC prior on the standard deviation parameter
of the likelihood:

```
# PC-prior for st. dev.
stdev.pcprior <- list(prior = "pc.prec", param = c(2, 0.01))
# Model fitting
res.barrier <- inla(form.barrier,
  data = inla.stack.data(joint.stk),
  control.predictor = list(A = inla.stack.A(joint.stk),
    compute = TRUE),
  family = 'gaussian',
  control.inla = list(int.strategy = "eb"),
  control.family = list(hyper = list(prec = stdev.pcprior)),
  control.mode = list(theta = c(1.647, 1.193, 3.975)))
```

The summary of the fitted model is shown below. Figure 5.11 shows estimates
of the posterior mean and 97.5% quantile of the noise. As can be seen, it is
clear that noise level at 1 am seems to be very close to or higher than the cut
off of 65 dB.

```
summary(res.barrier)
```
```
##
## Call:
##    c("inla(formula = form.barrier, family = \"gaussian\",
##    data = inla.stack.data(joint.stk), ", "
##    control.predictor = list(A = inla.stack.A(joint.stk),
##    compute = TRUE), ", " control.family = list(hyper =
##    list(prec = stdev.pcprior)), ", " control.inla =
##    list(int.strategy = \"eb\"), control.mode = list(theta
##    = c(1.647, ", " 1.193, 3.975)))")
## Time used:
##    Pre = 3.64, Running = 333, Post = 1.99, Total = 339
## Fixed effects:
##              mean     sd 0.025quant 0.5quant 0.975quant   mode kld
## intercept 64.98 1.505      62.03    64.98      67.93 64.98   0
##
## Random effects:
##    Name      Model
##      s RGeneric2
##
## Model hyperparameters:
##                                          mean     sd 0.025quant
## Precision for the Gaussian observations 8.08 8.735      1.076
## Theta1 for s                            1.21 0.210      0.802
## Theta2 for s                            3.98 0.396      3.212
##                                          0.5quant 0.975quant
## Precision for the Gaussian observations    5.48      30.82
## Theta1 for s                               1.20       1.63
## Theta2 for s                               3.98       4.77
##                                          mode
## Precision for the Gaussian observations 2.69
## Theta1 for s                            1.19
## Theta2 for s                            3.98
##
## Expected number of effective parameters(stdev): 6.93(0.00)
## Number of equivalent replicates : 1.01
##
## Marginal log-Likelihood:  -30.16
## Posterior marginals for the linear predictor and
##   the fitted values are computed
```

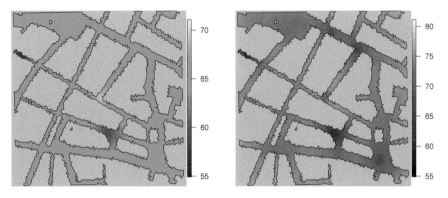

FIGURE 5.11 The left plot shows the posterior mean of the Barrier model, while the right plot shows the 97.5% quantile.

Space-time model

Space-time models are fully described in Chapter 7 and Chapter 8, but we have included here a simple example using the noise data. Given that information is available for 24 hours, it is possible to fit a space-time model to the noise data. First, a projector matrix is defined for the measurement points for the 24 hours:

```
A.st <- inla.spde.make.A(mesh = mesh,
  loc = coordinates(noise)[rep(1:7, 24), ])
```

The INLA stack is now created by adding variables LAeq1h, ... LAeq24h together with a time identifier (from 1 to 24 hours):

```
stk.st <- inla.stack(
  data = list(y = unlist(noise@data[, 2 + 1:24])),
  A = list(A.st, 1, 1),
  effects = list(s = 1:mesh$n,
    intercept = rep(1, 24 * nrow(noise)),
    time = rep(1:24, each = nrow(noise) )))
```

The model to be fitted is similar to the previous one, but now a cyclic random walk of order 1 is added as a separable time effect:

```
# Model formula
form.barrier.st <- y ~ 0 + intercept +
  f(s, model = barrier.model) +
```

```
f(time, model = "rw1", cyclic = TRUE, scale.model = TRUE,
  hyper = list(theta = stdev.pcprior))
```

Note that a PC prior has been used on the standard deviation parameter of the random walk effect. Then the model is fitted similarly as before:

```
res.barrier.st <- inla(form.barrier.st,
  data = inla.stack.data(stk.st),
  control.predictor = list(A = inla.stack.A(stk.st)),
  family = 'gaussian',
  control.inla = list(int.strategy = "eb"),
  control.family = list(hyper = list(prec = stdev.pcprior)),
  control.mode = list(theta = c(-1.883, 1.123, 3.995, -0.837)))
```

A summary of the fitted model can be seen below. Figure 5.12 shows noise estimates at 6, 12, 18 and 24 hours.

```
summary(res.barrier.st)
##
## Call:
##    c("inla(formula = form.barrier.st, family =
##    \"gaussian\", data = inla.stack.data(stk.st), ", "
##    control.predictor = list(A = inla.stack.A(stk.st)),
##    control.family = list(hyper = list(prec =
##    stdev.pcprior)), ", " control.inla = list(int.strategy
##    = \"eb\"), control.mode = list(theta = c(-1.883, ", "
##    1.123, 3.995, -0.837)))")
## Time used:
##     Pre = 3.28, Running = 526, Post = 0.273, Total = 530
## Fixed effects:
##                mean     sd 0.025quant 0.5quant 0.975quant    mode
## intercept -0.001 31.62      -62.08   -0.001      62.03 -0.001
##             kld
## intercept    0
##
## Random effects:
##   Name      Model
##      s RGeneric2
##   time RW1 model
##
## Model hyperparameters:
##                                            mean     sd 0.025quant
## Precision for the Gaussian observations 0.152 0.017      0.120
```

```
## Theta1 for s                                 1.137 0.253        0.647
## Theta2 for s                                 4.198 0.740        2.853
## Precision for time                           0.536 0.326        0.163
##                                              0.5quant 0.975quant
## Precision for the Gaussian observations       0.151       0.188
## Theta1 for s                                   1.134       1.641
## Theta2 for s                                   4.156       5.747
## Precision for time                            0.455       1.386
##                                               mode
## Precision for the Gaussian observations 0.150
## Theta1 for s                            1.123
## Theta2 for s                            4.002
## Precision for time                      0.336
##
## Expected number of effective parameters(stdev): 17.53(0.00)
## Number of equivalent replicates : 9.59
##
## Marginal log-Likelihood:  -443.29
```

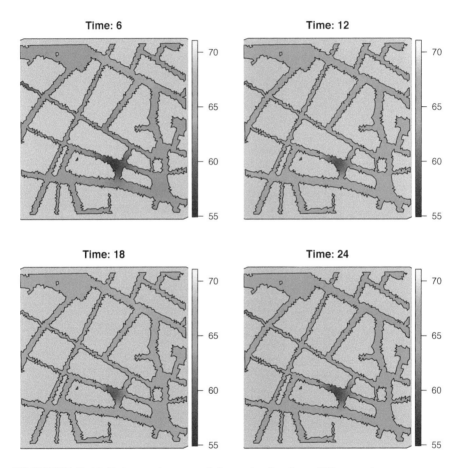

FIGURE 5.12 Point estimates of the noise level at different times.

6

Risk assessment using non-standard likelihoods

In this chapter we show how to use several different likelihoods that are related to risk assessment, combining them with the use of a spatial field.

The first example is about survival modeling where the outcome is the time to an event. This kind of outcome is common in medical studies where time to cure or time to death are common outcomes and in industry where time to failure is a common outcome. Most studies end before all of the patients are dead, or items have failed. In this case we use a censored likelihood; the likelihood considers the accumulated probability for the individual being alive until the study ends. We will only consider a simple case of censoring, but more complicated models are possible (see `?inla.surv`).

In the second example we consider modeling extreme events. In this case the data is, usually, the maximum over several observations collected over time, for example, annual maxima of daily rainfall. The common likelihood will not fit these kind of data and thus specific ones are considered. In the example in this chapter, the Generalized Extreme Value (GEV) distribution and the Pareto Distribution (PD) are being considered. We consider the inference for blockwise maxima data and the inference for threshold exceedances cases in order to illustrate two approaches for modeling this kind of data.

6.1 Survival analysis

In this section we show how to fit a survival model using a continuous spatial random effect modeled through the SPDE approach. The example is based on the data presented in Henderson et al. (2003). This data consists of 1043 cases of acute myeloid leukemia in adults recorded in New England between 1982 and 1998 by the North West Leukemia Register in the United Kingdom. The original code for the analysis is in Lindgren et al. (2011) and it has been adapted here to use the stack functionality. In Section 6.1.1 how to fit a parametric survival model is considered, while in Section 6.1.2 we show how to fit the semiparametric Cox proportional hazard model.

6.1.1   Parametric survival model

The Leuk dataset records the 1043 cases, and includes residential location and
other information about the patients. These are summarized in Table 6.1 and
full details are provided in Henderson et al. (2003).

TABLE 6.1: Description of the Leuk dataset on survival to acute
myeloid leukemia (AML).

Variable	Description
time	Survival time (in days).
cens	Death/censorship indicator (1 = observed, 0 = censored).
xcoord	x-coordinate of residence.
ycoord	y-coordinate of residence.
age	Age of the patient.
sex	Sex of the patient (0 = female, 1 = male).
wbc	White blood cell count (WBC) at diagnosis, truncated at 500 units.
tpi	Measure of deprivation for the enumeration district of residence (higher values indicate less affluent areas).
district	District of residence.

This is included in INLA and can be loaded and summarized as follows:

```
data(Leuk)
# Survival time as year
Leuk$time <- Leuk$time / 365
round(sapply(Leuk[, c(1, 2, 5:8)], summary), 2)
##            time cens   age  sex    wbc   tpi
## Min.       0.00 0.00 14.00 0.00   0.00 -6.09
## 1st Qu.    0.11 1.00 49.00 0.00   1.80 -2.70
## Median     0.51 1.00 65.00 1.00   7.90 -0.37
## Mean       1.46 0.84 60.73 0.52  38.59  0.34
## 3rd Qu.    1.47 1.00 74.00 1.00  38.65  2.93
## Max.      13.64 1.00 92.00 1.00 500.00  9.55
```

The reason we measure time in years instead of days is to standardize the
input for inla(). The likelihood for survival data has the exp(y * a) as its
expression, where a is a model parameter and y is the response. Thus, we must
be careful not to input too large or too small values, or the algorithm may
face numerical instabilities.

The Kaplan-Meyer maximum likelihood estimates of the survival curve by sex with the respective 95% confidence interval has been computed and visualized in Figure 6.1 with:

```
library(survival)
km <- survfit(Surv(time, cens) ~ sex, Leuk)
par(mar = c(2.5, 2.5, 0.5, 0.5), mgp = c(1.5, 0.5, 0), las = 1)
plot(km, conf.int = TRUE, col = 2:1)
legend('topright', c('female', 'male'), lty = 1, col = 2:1,
  bty = "n")
```

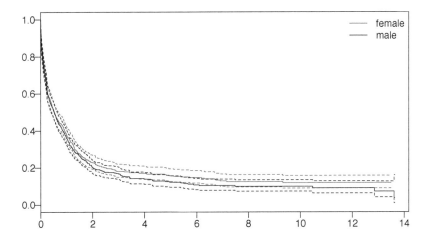

FIGURE 6.1 Survival time as function of gender.

The mesh for the SPDE model is built taking into account the coordinates available in the dataset, using the following code:

```
loc <- cbind(Leuk$xcoord, Leuk$ycoord)
nwseg <- inla.sp2segment(nwEngland)

bnd1 <- inla.nonconvex.hull(nwseg$loc, 0.03, 0.1, resol = 50)
bnd2 <- inla.nonconvex.hull(nwseg$loc, 0.25)
mesh <- inla.mesh.2d(loc, boundary = list(bnd1, bnd2),
  max.edge = c(0.05, 0.2), cutoff = 0.02)
```

Next, the projector matrix is obtained with:

```
A <- inla.spde.make.A(mesh, loc)
```

For the parameters of the SPDE model, namely the practical range and the marginal standard deviation, we will consider the PC-prior derived in Fuglstad et al. (2018), defined as:

```
spde <- inla.spde2.pcmatern(mesh = mesh,
  prior.range = c(0.05, 0.01), # P(range < 0.05) = 0.01
  prior.sigma = c(1, 0.01)) # P(sigma > 1) = 0.01
```

In this example, a `weibullsurv` likelihood is considered. However, other parametric likelihoods can be used as well; for example, up to date we have in `INLA`: `loglogistic`, `lognormal`, `exponential` and `weibullcure`. The Weibull distribution in `INLA` has two variants, see `inla.doc("weibull")`, and the default has a mean equal to $\Gamma(1 + 1/\alpha) \exp(-\alpha\eta)$, where α is a shape parameter and η is the linear predictor. Thus, the expected value is inversely proportional to the linear predictor, whereas the survival is directly proportional.

The formula for the linear predictor, including the intercept and covariates is

```
form0 <- inla.surv(time, cens) ~ 0 + a0 + sex + age + wbc + tpi
```

and adding the SPDE model is

```
form <- update(form0, . ~ . + f(spatial, model = spde))
```

Note how the formula includes a term using function `inla.surv()` to handle the observed time and the censoring status typical in survival data. This is common when working with survival outcomes due to the need to add the censoring information together with the survival time. The trick for building the data stack is to include all the variables needed in the formula. For the response these are `time` and the censoring status `cens`. The variables needed to define the spatial effect, namely the intercept and the covariates, are included as in previous models.

```
stk <- inla.stack(
  data = list(time = Leuk$time, cens = Leuk$cens),
  A = list(A, 1),
  effect = list(
    list(spatial = 1:spde$n.spde),
    data.frame(a0 = 1, Leuk[, -c(1:4)])))
```

Next, we will fit the model considering this data stack.

```
r <- inla(
```

```
form, family = "weibullsurv", data = inla.stack.data(stk),
control.predictor = list(A = inla.stack.A(stk), compute = TRUE))
```

Summary statistics of the posterior distribution of the intercept and the covariate effects can be extracted with the following code. Since we did not standardize the covariates, the effect sizes are not comparable.

```
round(r$summary.fixed, 4)
```

##		mean	sd	0.025quant	0.5quant	0.975quant	mode	kld
##	a0	-2.1718	0.2072	-2.5756	-2.1739	-1.7518	-2.1758	0
##	sex	0.0718	0.0692	-0.0641	0.0717	0.2076	0.0717	0
##	age	0.0327	0.0022	0.0284	0.0327	0.0371	0.0327	0
##	wbc	0.0031	0.0005	0.0021	0.0031	0.0039	0.0031	0
##	tpi	0.0245	0.0098	0.0051	0.0245	0.0437	0.0245	0

Similarly, summary statistics from the posterior distribution of the hyperparameters can be obtained as follows:

```
round(r$summary.hyperpar, 4)
```

##		mean	sd	0.025quant
##	alpha parameter for weibullsurv	0.5991	0.0160	0.5680
##	Range for spatial	0.3257	0.1637	0.1244
##	Stdev for spatial	0.2862	0.0719	0.1687

##		0.5quant	0.975quant	mode
##	alpha parameter for weibullsurv	0.5989	0.6310	0.5988
##	Range for spatial	0.2878	0.7466	0.2293
##	Stdev for spatial	0.2784	0.4500	0.2635

The spatial effect can be represented in a map. The polygons of each district in New England is also loaded when loading **Leuk**. First, a projection from the mesh into a grid is defined considering the bounding box around New England:

```
bbnw <- bbox(nwEngland)
r0 <- diff(range(bbnw[1, ])) / diff(range(bbnw[2, ]))
prj <- inla.mesh.projector(mesh, xlim = bbnw[1, ],
  ylim = bbnw[2, ], dims = c(200 * r0, 200))
```

Then, the spatial effect (i.e., posterior mean and standard deviation) is interpolated and NA values are assigned to all the grid points that are outside of the region of interest:

```
spat.m <- inla.mesh.project(prj, r$summary.random$spatial$mean)
spat.sd <- inla.mesh.project(prj, r$summary.random$spatial$sd)
ov <- over(SpatialPoints(prj$lattice$loc), nwEngland)
spat.sd[is.na(ov)] <- NA
spat.m[is.na(ov)] <- NA
```

The posterior mean and standard deviation are displayed in Figure 6.2. As a result, the spatial effect has continuous variation along the region, rather than being constant inside each district.

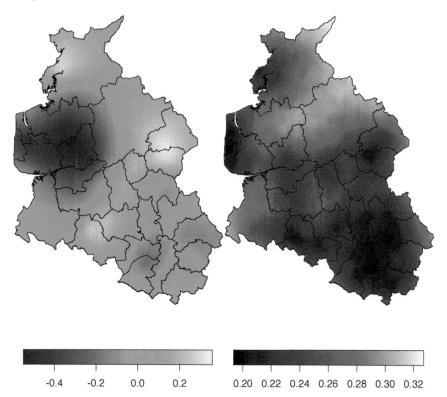

FIGURE 6.2 Map of the spatial effect for the Weibull survival model. Posterior mean (left) and posterior standard deviation (right).

6.1.2   Cox proportional hazard survival model

The Cox proportional hazard survival model (`coxph` family in INLA) is very common and we can fit it using maximum likelihood with the `coxph()` function from the `survival` package (Therneau, 2015; Therneau and Grambsch, 2000), with:

```
m0 <- coxph(Surv(time, cens) ~ sex + age + wbc + tpi, Leuk)
```

This model can be written as a Poisson regression. This idea was proposed in Holford (1980), detailed in Laird and Olivier (1981) and studied in Andersen and Gill (1982). In INLA we can prepare survival data for fitting a Cox proportional hazard model using the Poisson likelihood using the `inla.coxph()` function (Martino et al., 2010). We supply in this function a formula without the spatial effect, just to have the data prepared for fitting the Cox proportional hazard survival model as a Poisson model. The output from the `inla.coxph()` function will be supplied in the `inla.stack()` function together with the spatial terms.

```
cph.leuk <- inla.coxph(form0,
  data = data.frame(a0 = 1, Leuk[, 1:8]),
  control.hazard = list(n.intervals = 25))
```

For comparison purposes, we fit the model without the spatial effect considering the `coxph` family in INLA:

```
cph.res0 <- inla(form0, family = 'coxph',
  data = data.frame(a0 = 1, Leuk[, c(1,2, 5:8)]))
```

The next code changes the original formula updating it on the output and we will use it to add the spatial effect:

```
cph.formula <- update(cph.leuk$formula,
  '. ~ . + f(spatial, model = spde)')
```

The projector matrix can be built as follows:

```
cph.A <- inla.spde.make.A(mesh,
  loc = cbind(cph.leuk$data$xcoord, cph.leuk$data$ycoord))
```

Finally, the data stack is built considering the relevant data from the output from the `inla.coxph()` function:

```
cph.stk <- inla.stack(
  data = c(list(E = cph.leuk$E), cph.leuk$data[c('y..coxph')]),
  A = list(cph.A, 1),
  effects = list(
    list(spatial = 1:spde$n.spde),
```

```
        cph.leuk$data[c('baseline.hazard', 'a0',
            'age', 'sex', 'wbc', 'tpi')]))
```

```
cph.data <- c(inla.stack.data(cph.stk), cph.leuk$data.list)
```

Then, the model is fitted considering a Poisson likelihood:

```
cph.res <- inla(cph.formula, family = 'Poisson',
    data = cph.data, E = cph.data$E,
    control.predictor = list(A = inla.stack.A(cph.stk)))
```

We now compare the estimated fixed effects from these results:

```
round(data.frame(surv = coef(summary(m0))[, c(1,3)],
    r0 = cph.res0$summary.fixed[-1, 1:2],
    r1 = cph.res$summary.fixed[-1, 1:2]), 4)
##      surv.coef surv.se.coef  r0.mean  r0.sd r1.mean  r1.sd
## sex   0.0522      0.0678      0.0579 0.0679  0.0685 0.0692
## age   0.0296      0.0021      0.0333 0.0021  0.0348 0.0023
## wbc   0.0031      0.0004      0.0034 0.0005  0.0034 0.0005
## tpi   0.0293      0.0090      0.0342 0.0090  0.0322 0.0098
```

Regarding the spatial effects, the fitted values of the spatial effects are very similar between the two models fitted (Weibull and Cox):

```
s.m <- inla.mesh.project(prj, cph.res$summary.random$spatial$mean)
cor(as.vector(spat.m),  as.vector(s.m), use = 'p')
## [1] 0.9942
s.sd <- inla.mesh.project(prj, cph.res$summary.random$spatial$sd)
cor(log(as.vector(spat.sd)), log(as.vector(s.sd)), use = 'p')
## [1] 0.9987
```

6.2 Models for extremes

6.2.1 Motivation

Extreme Value Theory (EVT) emerges as the natural tool to assess probabilities of events that are rare, and whose occurrence involves great risk (Coles, 2001). From the applied point of view, EVT provides a framework to develop techniques and models to address important problems related to risk assessment in many different areas, such as hydrology, wind engineering, climate change, flood monitoring and prediction, and large insurance claims.

When modeling extreme events, we typically need to extrapolate beyond observed data, into the tail of the distribution, where the available data are often limited, and classical inference methods usually fail. The plausibility of the extrapolation is subject to certain stability conditions that constrain the class of possible distributions on which extrapolation should be based.

In the context of univariate observations, EVT is based on an asymptotic characterization of maxima of a collection of independent and identically distributed (i.i.d.) random variables. Given i.i.d. continuous data $Y_1, \ldots, Y_n$ and $M_n = \max\{Y_1, \ldots, Y_n\}$, it can be shown that if the normalized distribution of M_n converges as $n \to \infty$, then it converges to a Generalized Extreme-Value (GEV) distribution. The GEV is characterized by a location, a scale, and shape parameters (see, e.g., Coles, 2001, Chapter 3). In practice, this asymptotic result serves as a justification to use the GEV distribution to model maxima over finite blocks of time, like monthly or annual maxima. Nevertheless, if the data are available at a finer resolution, then an alternative characterization of extremes might be given by the so-called *threshold exceedances* approach, where the interest is on observations that exceed a certain high predefined threshold. It can be shown that if the conditions for the GEV limit of normalized maxima hold, then as the threshold becomes large, the distribution of threshold exceedances converges to the generalized Pareto (GP) distribution characterized by a scale and shape parameters.

In practical applications, extreme observations are rarely independent. This has motivated the development of methodology to account for temporal non-stationarity in the distributions of interest. Typical approaches to this problem are based on the generalized linear modeling framework, allowing the parameters of a marginal distribution to depend on covariates (Davison and Smith, 1990). Semi-parametric approaches for block maxima of bivariate vectors were introduced by Castro-Camilo and de Carvalho (2017) and Castro-Camilo et al. (2018). Hierarchical Bayesian models for threshold exceedances were introduced in Casson and Coles (1999), Cooley et al. (2007), and Opitz et al. (2018). Here we assume that GEV and GP observations are conditionally independent given a latent process that drives the spatial (or space-time) trends, as well as the dependence structure. In the following sections, we show how to simulate from and fit such models under the SPDE framework.

6.2.2   Simulating from the GEV and GP distributions

The random field and the linear predictor

We start by generating $n = 200$ random locations where we assume data are observed. These locations are also used as triangulation nodes to create the mesh:

```
library(INLA)
set.seed(1)
n <- 200
loc.data <- matrix(runif(n * 2), n, 2)
mesh <- inla.mesh.2d(loc = loc.data, cutoff = 0.05,
  offset = c(0.1, 0.4), max.edge = c(0.05, 0.5))
```

The selected cutoff avoids creating too many small triangles. The SPDE model defined at the nodes of the mesh is constructed as follows:

```
spde <- inla.spde2.pcmatern(mesh,
  prior.range = c(0.5, 0.5),
  prior.sigma = c(0.5, 0.5))
```

Note that we use a PC-prior specification for the range and the marginal standard deviation. Specifically, we assume that P(range $< 0.5) = 0.5$ and P(std. dev $> 0.5) = 0.5$. We compute the precision matrix (Qu) for the true range (range) and marginal standard deviation (sigma.u) as follows:

```
sigma.u <- 2
range <- 0.7
Qu <- inla.spde.precision(spde,
  theta = c(log(range), log(sigma.u)))
```

Now we can generate $m = 40$ samples from the spatial field **u**:

```
m <- 40 # number of replications in time
set.seed(1)
u <- inla.qsample(n = m, Q = Qu, seed = 1)
```

The spatial field **u** is a continuous simulation for the entire study area. To obtain simulations over the n random locations previously generated, we project **u** over the locations:

```
A <- inla.spde.make.A(mesh = mesh, loc = loc.data)
u <- (A %*% u)
```

Note that the dimension of `A` is:

```
dim(A)
## [1] 200 462
```

This is the number of locations (n) times the number of nodes in the mesh (`mesh$n`). We are now able to define our linear predictor. For simplicity, we choose a linear predictor of the form $\eta_i = \beta_0 + \beta_1 x + \mathbf{u}(\mathbf{s}_i)$, where x is a covariate and $\mathbf{u}(\mathbf{s}_i)$ is the spatial field at location $\mathbf{s}_i$:

```
b_0 <- 1 # intercept
b_1 <- 2 # coefficient for covariate
set.seed(1)
covariate <- rnorm(m*n)
lin.pred <- b_0 + b_1 * covariate + as.vector(u)
```

Samples from the GEV distribution

In INLA, the GEV distribution is parametrized in terms of a location parameter $\mu \in \mathbb{R}$, a shape parameter $\xi \in \mathbb{R}$, and a precision parameter $\tau > 0$:

$$G(y; \mu, \tau, \xi) = \exp\left\{-\left[1 + \xi\sqrt{\tau s}(y - \mu)\right]^{-1/\xi}\right\}, \quad \text{for } 1 + \xi\sqrt{\tau s}(y - \mu) > 0, \tag{6.1}$$

where $s > 0$ is a fixed scaling. The current implementation only allows the linear predictor to be linked to the location parameter using the identity link, i.e., $\mu = \eta$. We generate $m \times n$ samples from a GEV distribution using the `rgev()` function from the `evd` package (Stephenson, 2002) as follows:

```
s <- 0.01
tau <- 4
s.y <- 1/sqrt(s*tau) # true scale
xi.gev <- 0.1 # true shape
library(evd)
set.seed(1)
y.gev <- rgev(n = length(lin.pred), loc = lin.pred,
   shape = xi.gev, scale = s.y)
```

Samples from the GP distribution

The generalized Pareto distribution has a cumulative distribution function given by

$$G(y; \sigma, \xi) = 1 - \left(1 + \xi \frac{y}{\sigma}\right)^{-1/\xi}, \quad y > 0, \ \xi \neq 0.$$

Note that the exponential distribution arises in the limiting case, as for $\xi \to 0$, we have $G(y; \sigma) = 1 - \exp(-y/\sigma)$. In INLA, the linear predictor controls the α-quantile of the GP distribution:

$$P(y \leq q_\alpha) = \alpha,$$
$$q_\alpha = \exp(\eta),$$

where $\alpha \in (0, 1)$ is provided by the user. The scale parameter $\sigma \in \mathbb{R}$ is then a function of q_α and ξ:

$$\sigma = \frac{\xi \exp(\eta)}{(1 - \alpha)^{-\xi} - 1}.$$

Using the probability integral transform, we can generate samples from the GP distribution by defining the following function:

```
rgp = function(n, sigma, eta, alpha, xi = 0.001){
  if (missing(sigma)) {
    stopifnot(!missing(eta) && !missing(alpha))
    sigma = exp(eta) * xi / ((1.0 - alpha)^(-xi) - 1.0)
  }
  return (sigma / xi * (runif(n)^(-xi) - 1.0))
}
```

Note that function `rgp()` is conveniently parametrized in terms of the scale σ, the linear predictor η, and the probability α (i.e., the INLA parametrization). A sample of size $m \times n$ from the GP can then be obtained as follows:

```
xi.gp <- 0.3
alpha <- 0.5 # median
q <- exp(lin.pred)
scale <- xi.gp * q / ((1 - alpha)^(-xi.gp) - 1)
set.seed(1)
y.gp <- rgp(length(lin.pred), sigma = scale, eta = lin.pred,
  alpha = alpha, xi = xi.gp)
```

Note that in this case, the linear predictor is linked to the median of the distribution (since $\alpha = 0.5$).

6.2.3   Inference for the GEV and GP distributions

Inference for blockwise maxima

We start by generating a new mesh and A matrix to fit our model:

```
mesh <- inla.mesh.2d(loc = loc.data, cutoff = 0.05,
  max.edge = c(0.05, 0.1))
rep.id <- rep(1:m, each = n)
x.loc <- rep(loc.data[, 1], m)
y.loc <- rep(loc.data[, 2], m)
A <- inla.spde.make.A(mesh = mesh, loc = cbind(x.loc, y.loc),
  group = rep.id)
```

Note that the dimension of the matrix A (8000, 24960) that we use for inference is much bigger that the one we used before. This is because we have $m = 40$ replications in time of a spatial field projected over $n = 200$ locations ($n \times m = 8000$), and the dimension of our new mesh is mesh$n = 624 ($m \times$ mesh$n = 24960$).

We specify our prior model for the spatial effect, on the median of the parameters, as follows:

```
prior.median.sd <- 1
prior.median.range <- 0.7
spde <- inla.spde2.pcmatern(mesh,
  prior.range = c(prior.median.range, 0.5),
  prior.sigma = c(prior.median.sd, 0.5))
```

We obtain the named index vectors required for the SPDE model:

```
mesh.index <- inla.spde.make.index(name = "field",
  n.spde = spde$n.spde, n.group = m)
```

We stack our data using the inla.stack() function:

```
stack <- inla.stack(
  data = list(y = y.gev),
  A = list(A, 1, 1),
  effects = list(mesh.index,
```

```
      intercept = rep(1, length(lin.pred)),
      covar = covariate),
    tag = "est")
```

Posterior marginals for the parameters of the GEV model are obtained by fitting the model with INLA. Here we use initial values to speed up inference (using variable init below). If you do not know what initial values to use, init can be set to NULL.

```
# Model formula
formula <- y ~ -1 + intercept + covar +
  f(field, model = spde, group = field.group,
    control.group = list(model = "iid"))
# Initial values of hyperparameters
init = c(-3.253,  9.714, -0.503,  0.595)
# Prior on GEV parameters
hyper.gev = list(theta2 = list(prior = "gaussian",
  param = c(0, 0.01), initial = log(1)))
# Mode fitting
res.gev <- inla(formula,
  data = inla.stack.data(stack),
  family ="gev",
  control.inla = list(strategy = "adaptive"),
  control.mode = list(restart = TRUE, theta = init),
  control.family = list(hyper = hyper.gev,
    gev.scale.xi = 0.01),
  control.predictor = list(A = inla.stack.A(stack),
    compute = TRUE))
```

We specify model = 'iid' because we are assuming to have independent replicates in time. A summary for the model hyperparameters, i.e., the practical range and the marginal standard deviation for the latent Gaussian random field, along with the scale and shape parameters of the GEV can be obtained as:

```
tab.res.GEV <- round(cbind(true = c(1 / s.y^2, xi.gev, range,
    sigma.u),
  res.gev$summary.hyperpar), 4)
```

This summary is available in Table 6.2 together with the true values of the parameters used to simulate the data.

TABLE 6.2: Summary of the posterior distributions of the parameters and their true values.

Parameter	True	Mean	St. Dev.	2.5% quant.	97.5% quant.
Precision for GEV	0.04	0.0389	0.0009	0.0372	0.0407
Shape for GEV	0.10	0.0994	0.0094	0.0810	0.1179
Range for field	0.70	0.6079	0.0590	0.4991	0.7314
Stdev for field	2.00	1.8786	0.0989	1.6927	2.0811

Inference for threshold exceedances

The shape parameter $\xi \in \mathbb{R}$ plays an important role in determining the weight of the upper tail: heavy tails are obtained with $\xi > 0$, light tails with $\xi \to 0$, and bounded tails with $\xi < 0$. Moreover, if $\xi \geq 1$ then the mean is infinite, and if $\xi \geq 1/2$ then the variance is infinite. Therefore, for many practical applications, large values of ξ are unrealistic. Here we show how to specify this prior information using an approximation to the PC-prior for ξ.

The current implementation for the GP distribution in `INLA` allows us to specify an Gamma prior distribution for ξ (or in the `INLA` parametrization, log-Gamma for $\theta = \log(\xi)$) that resembles a PC-prior for suitable choices of the Gamma shape and rate parameters. Specifically, Opitz et al. (2018) showed that the PC-prior for ξ when the simpler model is the exponential distribution (i.e. $\xi = 0$) is

$$\pi(\xi) = \tilde{\lambda} \exp\left\{-\tilde{\lambda}\frac{\xi}{(1-\xi)^{1/2}}\right\}\left\{\frac{1-\xi/2}{(1-\xi)^{3/2}}\right\}, \quad 0 \leq \xi < 1,$$

where $\lambda = \tilde{\lambda}/\sqrt{2}$ is the penalization rate parameter (Fuglstad et al., 2018). The PC-prior $\pi(\xi)$ can be approximated when $\xi \to 0$ by an exponential distribution with rate $\tilde{\lambda}$ (or Gamma distribution with unit shape and rate $\tilde{\lambda}$):

$$\tilde{\pi}(\xi) = \sqrt{2}\lambda \exp(-\sqrt{2}\lambda\xi) = \tilde{\lambda}\exp(-\tilde{\lambda}\xi), \quad \xi \geq 0.$$

Figure 6.3 shows that both priors are very similar to each other for large values of the penalization rate λ, and differ for smaller values.

As mentioned before, we would like to specify a prior for ξ that drifts away from large values. To do this, we choose $\lambda = 10$, which implies that $P(\xi > 0.2) = 0.06$ under $\tilde{\pi}$ (see Figure 6.3). We specify this in the `INLA` framework as follows:

```
hyper.gp = list(theta = list(prior = "loggamma",
  param = c(1, 10)))
```

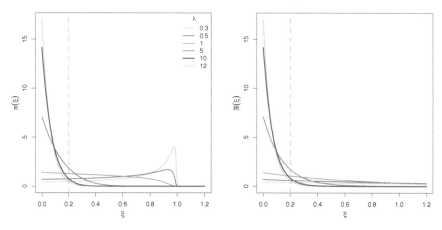

FIGURE 6.3 PC-priors for the GP shape parameter ξ for different values of the penalization rate λ.

Using the SPDE model defined in Section 6.2.3, we fit our model for exceedances as follows:

```
# Data stack
stack <- inla.stack(
  data = list(y = y.gp),
  A = list(A, 1, 1),
  effects = list(mesh.index,
    intercept = rep(1, length(lin.pred)),
    covar = covariate),
  tag = "est")
#Initial values of the hyperparameters
init2 <- c(-1.3, -0.42, 0.62)
# Model fitting
res.gp <- inla(formula,
  data = inla.stack.data(stack, spde = spde),
  family ="gp",
  control.inla = list(strategy = "adaptive"),
  control.mode = list(restart = TRUE, theta = init2),
  control.family = list(list(control.link = list(quantile = 0.5),
    hyper = hyper.gp)),
  control.predictor = list(A = inla.stack.A(stack),
    compute = TRUE))
```

As with the GEV fit, the summary for the model hyperparameters can be obtained as:

```
table.results.GP <- round(cbind(true = c(xi.gp, range, sigma.u),
            res.gp$summary.hyperpar), 4)
```

This is available in Table 6.3, together with the true values of the parameters.

TABLE 6.3: Summary of the posterior distributions.

Parameter	True	Mean	St. Dev.	2.5% quant.	97.5% quant.
Shape for GP	0.3	0.2711	0.0212	0.2308	0.3141
Range for field	0.7	0.6596	0.0317	0.6003	0.7246
Stdev for field	2.0	1.8616	0.0611	1.7456	1.9855

7

Space-time models

In this chapter we detail how to fit separable space-time models. A separable space-time model is defined as a SPDE model for the spatial domain and an autoregressive model of order 1, i.e., AR(1), for the time dimension. The space-time separable model is defined by the Kronecker product between the precision matrices of the spatial and temporal random effects. Additional information about separable space-time models can be found in Cameletti et al. (2013).

In this chapter we start by showing two different ways to implement space-time models. The first one uses discrete time domain and the second one considers continuous time and discretizes this over a set of knots. The main difference in the model fitting process is that when we use continuous time, we need to choose time knots and to adjust the projector matrix to use these knots. However, none of the approaches requires the measurement locations to be the same over time.

In this chapter we focus on basic code examples, together with information on how to structure the models for faster computation. In the following chapter we will provide several advanced examples.

7.1 Discrete time domain

In this section we show how to fit a space-time separable model, as in Cameletti et al. (2013). Additionally, we show how to include a categorical covariate.

7.1.1 Data simulation

The study region considered in this example is the border of Paraná state, available in package `INLA`, as in Section 2.8. This boundary will be used as the domain of the spatial process and it can be loaded as:

```
data(PRborder)
```

The first step is to define the spatial model. To be able to fit the model quickly, we use the low resolution mesh for Paraná state border created in Section 2.6.

There are two options to simulate from the model proposed in Cameletti et al. (2013). The first one is based on the simultaneous distribution of the latent field and the second one is based on the conditional simulation at each time. This last option is easy to compute as each time point is simulated conditionally on the previous one, giving linear run time for long temporal simulations.

First, the time dimension is set to $k = 12$:

```
k <- 12
```

The locations of the points are the same as in dataset PRprec, but considered in a randomized order:

```
data(PRprec)
coords <- as.matrix(PRprec[sample(1:nrow(PRprec)), 1:2])
```

In the following simulation step we will use the book.rspde() function available in the file spde-book-functions.R. The k independent realizations of the spatial model can be generated as follows:

```
params <- c(variance = 1, kappa = 1)

set.seed(1)
x.k <- book.rspde(coords, range = sqrt(8) / params[2],
  sigma = sqrt(params[1]), n = k, mesh = prmesh1,
  return.attributes = TRUE)
```

The number of space-time observations is the number of rows of x.k and it can be checked with:

```
dim(x.k)
## [1] 616   12
```

Now, the autoregressive parameter ρ for the temporal effect is defined:

```
rho <- 0.7
```

Next, temporal correlation is introduced:

```
x <- x.k
for (j in 2:k)
  x[, j] <- rho * x[, j - 1] + sqrt(1 - rho^2) * x.k[, j]
```

Here, the $\sqrt{1 - \rho^2}$ term is added to make the process stationary in time, see Rue and Held (2005) and Cameletti et al. (2013). Figure 7.1 shows the realization of the space-time process.

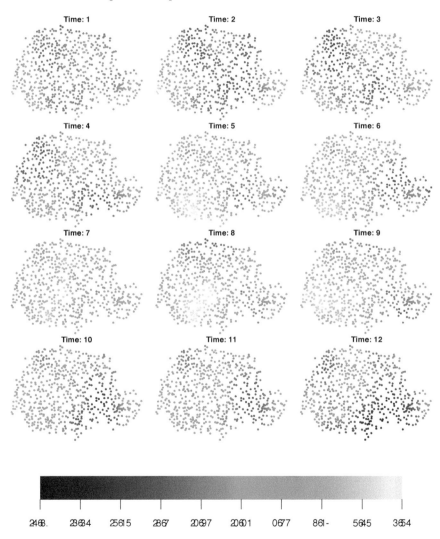

FIGURE 7.1 Realization of the space-time random field.

In this example, a categorical covariate will be included in the model. We simulate a categorical covariate with three levels (labeled A, B and C):

```
n <- nrow(coords)
set.seed(2)
ccov <- factor(sample(LETTERS[1:3], n * k, replace = TRUE))
```

The distribution of values of this categorical covariate is:

```
table(ccov)
## ccov
##    A    B    C
## 2458 2438 2496
```

The regression coefficients and the regression parameters are:

```
beta <- -1:1
```

The response variable will be computed by adding the fixed effect on the categorical covariate, the spatio-temporal random effect and some random white noise (with standard deviation 0.1):

```
sd.y <- 0.1
y <- beta[unclass(ccov)] + x + rnorm(n * k, 0, sd.y)
```

The average value of the response on the levels of the categorical covariate are:

```
tapply(y, ccov, mean)
##         A          B          C
## -1.09946 -0.09181    0.91967
```

To show that we can use different locations at different times, some of the observations will be dropped. In particular, only half of the simulated data will be kept. This can be done by creating an index for the selected observations, as follows:

```
isel <- sample(1:(n * k), n * k / 2)
```

These data are then put together in a data.frame:

```
dat <- data.frame(y = as.vector(y), w = ccov,
  time = rep(1:k, each = n),
  xcoo = rep(coords[, 1], k),
  ycoo = rep(coords[, 2], k))[isel, ]
```

In real applications there may be completely misaligned locations across different times. The code we provide in this example will work in that situation.

7.1.2 Data stack preparation

We use the PC-priors derived in Fuglstad et al. (2018) for the model parameters range and marginal standard deviation. These are set when defining the SPDE, as follows:

```
spde <- inla.spde2.pcmatern(mesh = prmesh1,
  prior.range = c(0.5, 0.01), # P(range < 0.05) = 0.01
  prior.sigma = c(1, 0.01)) # P(sigma > 1) = 0.01
```

Now, additional data preparation is required to build the space-time model. The index set is made taking into account the number of mesh points in the SPDE model and the number of groups, as:

```
iset <- inla.spde.make.index('i', n.spde = spde$n.spde,
  n.group = k)
```

Note that the index set for the latent field does not depend on the data set locations. It only depends on the SPDE model size and on the time dimension. The projection matrix is defined using the coordinates of the observed data. We need to pass the time index to the **group** argument to build the projector matrix and the **inla.spde.make.A()** function:

```
A <- inla.spde.make.A(mesh = prmesh1,
  loc = cbind(dat$xcoo, dat$ycoo), group = dat$time)
```

The **effects** in the stack is a list with two elements: the first one is the index set and the second one the categorical covariate. The stack data is defined as:

```
sdat <- inla.stack(
  data = list(y = dat$y),
  A = list(A, 1),
  effects = list(iset, w = dat$w),
  tag = 'stdata')
```

7.1.3   Fitting the model and some results

In this example, a PC-prior (see Section 1.6.5) is also used for the temporal autoregressive parameter, i.e., the autocorrelation parameter. In particular, this prior considers that $P(cor > 0) = 0.9$ and it is defined as follows:

```
h.spec <- list(theta = list(prior = 'pccor1', param = c(0, 0.9)))
```

To deal with the categorical covariate we need to use `expand.factor.strategy = 'inla'` in the `control.fixed` argument list to get an intuitive result. Hence, model fitting is done as follows:

```
# Model formula
formulae <- y ~ 0 + w + f(i, model = spde, group = i.group,
  control.group = list(model = 'ar1', hyper = h.spec))
# PC prior on the autoreg. param.
prec.prior <- list(prior = 'pc.prec', param = c(1, 0.01))
# Model fitting
res <- inla(formulae,  data = inla.stack.data(sdat),
  control.predictor = list(compute = TRUE,
    A = inla.stack.A(sdat)),
  control.family = list(hyper = list(theta = prec.prior)),
  control.fixed = list(expand.factor.strategy = 'inla'))
```

A summary of the three intercepts, together with the observed mean for each covariate level, is:

```
cbind(Obs. = tapply(dat$y, dat$w, mean), res$summary.fixed[, -7])
##        Obs.    mean    sd 0.025quant 0.5quant 0.975quant    mode
## A -1.0851 -1.324 0.438     -2.193   -1.323     -0.461 -1.321
## B -0.0785 -0.318 0.438     -1.187   -0.317      0.545 -0.315
## C  0.9014  0.681 0.438     -0.188    0.682      1.544  0.684
```

The posterior marginal distributions for the random field parameters and the marginal distribution for the temporal correlation are displayed in Figure 7.2.

7.1.4   A look at the posterior random field

The random field posterior distribution can be compared to the realized random field by means of the posterior mean, median, mode or any other quantile.

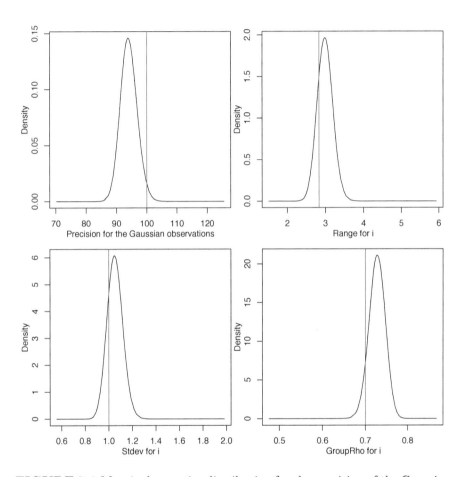

FIGURE 7.2 Marginal posterior distribution for the precision of the Gaussian likelihood (top-left), the practical range (top-right), standard deviation of the field (bottom-left) and the temporal correlation (bottom-right). The red vertical lines are placed at the true values of the parameters.

Before we get to this point, we need the index for the random field at the data locations:

```
idat <- inla.stack.index(sdat, 'stdata')$data
```

The correlation between the simulated data response and the posterior mean of the predicted values can be computed as follows:

```
cor(dat$y, res$summary.linear.predictor$mean[idat])
## [1] 0.9982
```

The correlation is almost one because there is no error term in the model.

We now compute predictions for each time point. First, a grid is defined in the same way as in the rainfall example in Section 2.8:

```
stepsize <- 4 * 1 / 111
nxy <- round(c(diff(range(coords[, 1])),
  diff(range(coords[, 2]))) / stepsize)
projgrid <- inla.mesh.projector(
  prmesh1, xlim = range(coords[, 1]),
  ylim = range(coords[, 2]), dims = nxy)
```

Then, the prediction for each time can be done as follows:

```
xmean <- list()
for (j in 1:k)
  xmean[[j]] <- inla.mesh.project(
    projgrid, res$summary.random$i$mean[iset$i.group == j])
```

Next, we subset to the points of the grid inside the boundaries of Paraná state, and set the points of the grid out of the Paraná border to NA:

```
library(splancs)
xy.in <- inout(projgrid$lattice$loc,
  cbind(PRborder[, 1], PRborder[, 2]))
```

We visualize the result in Figure 7.3.

7.1.5 Validation

The inference results we just showed are based only on part of the simulated data. The other part of the simulated data can now be used for validation.

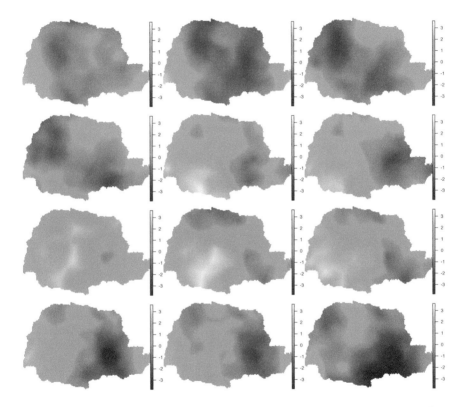

FIGURE 7.3 Visualization of the posterior mean of the space-time random field. Time flows from top to bottom and left to right.

Therefore, another data stack is required to compute posterior distributions for the validation data:

```
vdat <- data.frame(r = as.vector(y), w = ccov,
  t = rep(1:k, each = n), x = rep(coords[, 1], k),
  y = rep(coords[, 2], k))
vdat <- vdat[-isel, ]
```

We compute a projection matrix and the data stack for the validation data:

```
Aval <- inla.spde.make.A(prmesh1,
  loc = cbind(vdat$x, vdat$y), group = vdat$t)
stval <- inla.stack(
  data = list(y = NA), # NA: no data, only enable predictions
  A = list(Aval, 1),
  effects = list(iset, w = vdat$w),
  tag = 'stval')
```

Next, we join these two stacks together into a full stack and re-fit the model. We use the estimates of the hyperparameters obtained with the previous model to speed up computations:

```
stfull <- inla.stack(sdat, stval)
vres <- inla(formulae,  data = inla.stack.data(stfull),
  control.predictor = list(compute = TRUE,
    A = inla.stack.A(stfull)),
  control.family = list(hyper = list(theta = prec.prior)),
  control.fixed = list(expand.factor.strategy = 'inla'),
  control.mode = list(theta = res$mode$theta, restart = FALSE))
```

Predicted values versus observed values have be plotted in Figure 7.4 to assess goodness of fit; they are in close agreement. The indices of the fitted values to be used when extracting the results from the `inla` object for plotting have been obtained with:

```
ival <- inla.stack.index(stfull, 'stval')$data
```

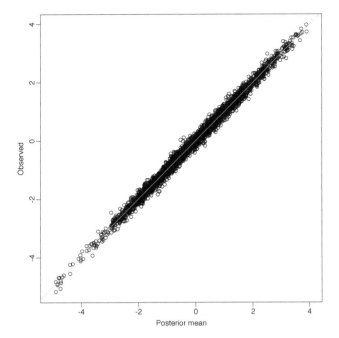

FIGURE 7.4 Validation: Observed values versus posterior means from the fitted model.

7.2 Continuous time domain

We now eliminate the assumption that the observations have been collected over discrete time points. This is the case for, e.g., fishing data, and space-time point processes in general. Similarly to how we use the Finite Element Method approach for space, we use a set of time knots to set up piecewise linear basis functions over time.

7.2.1 Data simulation

First, we set the spatial locations and sample time points from a continuous interval:

```
loc <- unique(as.matrix(PRprec[, 1:2]))
n <- nrow(loc)
time <- sort(runif(n, 0, 1))
```

To sample from the current model, we define a space-time separable covariance function. We use a Matérn covariance in space and exponential decaying covariance function over time:

```
local.stcov <- function(coords, time, kappa.s, kappa.t,
  variance = 1, nu = 1) {
  s <- as.matrix(dist(coords))
  t <- as.matrix(dist(time))
  scorr <- exp((1 - nu) * log(2) + nu * log(s * kappa.s) -
    lgamma(nu)) * besselK(s * kappa.s, nu)
  diag(scorr) <- 1
  return(variance * scorr * exp(-t * kappa.t))
}
```

Function `local.stcov()` will be used to compute the covariance function at the simulated space-time points and to sample from the model:

```
kappa.s <- 1
kappa.t <- 5
s2 <- 1 / 2
xx <- crossprod(
  chol(local.stcov(loc, time, kappa.s, kappa.t, s2)),
  rnorm(n))

beta0 <- -3
tau.error <- 3

y <- beta0 + xx + rnorm(n, 0, sqrt(1 / tau.error))
```

7.2.2 Data stack preparation

To fit the space-time continuous model we must first define the time knots and the temporal mesh. For this, we define a one-dimensional mesh with 10 knots:

```
k <- 10
mesh.t <- inla.mesh.1d(seq(0 + 0.5 / k, 1 - 0.5 / k, length = k))
```

The knots in the resulting temporal mesh are the following:

```
mesh.t$loc
##  [1] 0.05 0.15 0.25 0.35 0.45 0.55 0.65 0.75 0.85 0.95
```

We continue using the low resolution mesh for the border of Paraná state

created in Section 2.6. This means that we can also re-use the SPDE model
defined in the previous example.

The index set for the spatio-temporal model can be defined as:

```
iset <- inla.spde.make.index('i', n.spde = spde$n.spde,
  n.group = k)
```

The projection matrix considers both the spatial and temporal projection.
Hence, it needs the spatial mesh and the spatial locations, the time points
and the temporal mesh. These are passed to function `inla.spde.make.A` as
follows:

```
A <- inla.spde.make.A(mesh = prmesh1, loc = loc,
  group = time, group.mesh = mesh.t)
```

The `effects` in the data stack are a list with two elements:
the index set for the spatial effect and the categorical covariate. The stack
data is defined as:

```
sdat <- inla.stack(
  data = list(y = y),
  A = list(A,1),
  effects = list(iset, list(b0 = rep(1, n))),
  tag = "stdata")
```

7.2.3 Fitting the model and some results

An exponential correlation function is used for time with parameter κ as the
inverse range parameter. It gives a correlation between time knots equal to:

```
exp(-kappa.t * diff(mesh.t$loc[1:2]))
## [1] 0.6065
```

We fit the model using an AR(1) temporal correlation over the time knots as
follows:

```
formulae <- y ~ 0 + b0 + f(i, model = spde, group = i.group,
  control.group = list(model = 'ar1', hyper = h.spec))

res <- inla(formulae, data = inla.stack.data(sdat),
  control.family = list(hyper = list(theta = prec.prior)),
```

```
control.predictor = list(compute = TRUE,
  A = inla.stack.A(sdat)))
```

We summarize the posterior marginal distributions for the likelihood precision and the random field parameters:

```
res$summary.hyperpar
##                                                mean        sd
## Precision for the Gaussian observations  2.8535  0.19028
## Range for i                              2.4526  0.43626
## Stdev for i                              0.6753  0.07249
## GroupRho for i                           0.5124  0.14638
##                                          0.025quant  0.5quant
## Precision for the Gaussian observations      2.4918    2.8496
## Range for i                                  1.7246    2.4069
## Stdev for i                                  0.5423    0.6721
## GroupRho for i                               0.1879    0.5270
##                                          0.975quant     mode
## Precision for the Gaussian observations      3.2401  2.8449
## Range for i                                  3.4360  2.3132
## Stdev for i                                  0.8274  0.6664
## GroupRho for i                               0.7559  0.5585
```

The posterior marginal distributions of these parameters are shown in Figure 7.5, which includes the marginal distribution for the intercept, error precision, spatial range, standard deviation and temporal correlation in the space-time field.

7.3   Lowering the resolution of a spatio-temporal model

Model fitting can be challenging when dealing with large data sets. In this section we show techniques for lowering the resolution of the representation of the space-time random effect, to make model fitting faster.

First, we build the spatial mesh and the SPDE model using the rainfall data in Paraná state with the following code:

```
data(PRprec)
bound <- inla.nonconvex.hull(as.matrix(PRprec[, 1:2]), 0.2, 0.2,
  resol = 50)
```

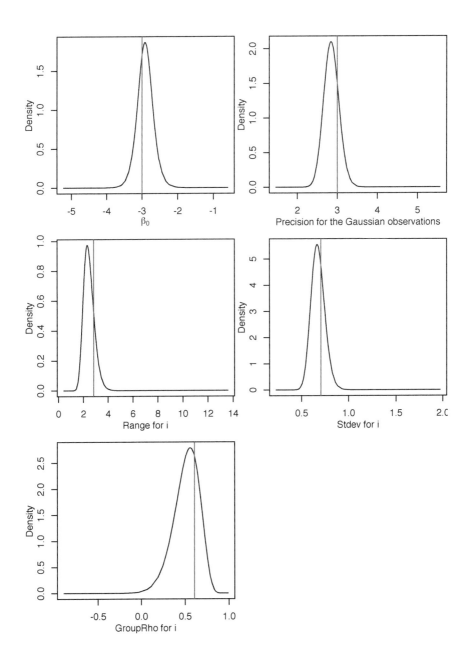

FIGURE 7.5 Marginal posterior distribution for the intercept, likelihood precision and the parameters in the space-time process.

```
mesh.s <- inla.mesh.2d(bound = bound, max.edge = c(1,2),
  offset = c(1e-5, 0.7), cutoff = 0.5)
spde.s <- inla.spde2.matern(mesh.s)
```

7.3.1   Data temporal aggregation

In this subsection we set up the data that we will use for the example in the next subsection. The reader may consider the dataframe df as an original binomial dataset, and how it was constructed is not of any large importance.

The data we analyze are composed of 616 location points observed over 365 days in Paraná state (Brazil). The response variable is daily rainfall. The dimension of the data.frame with this dataset and the first 7 variables from the first two rows:

```
dim(PRprec)
## [1] 616 368
PRprec[2:3, 1:7]
##    Longitude Latitude Altitude d0101 d0102 d0103 d0104
## 3    -50.77   -22.96     344     0     1     0    0.0
## 4    -50.65   -22.95     904     0     0     0    3.3
```

In this example the aim is to analyze the probability of rain. Therefore we now convert this continuous dataset of rainfall amount into occurrence of rain. The response variable is whether rainfall was higher than 0.1 or not.

```
PRoccurrence = 0 + (PRprec[, -c(1, 2, 3)] > 0.1)
PRoccurrence[2:3, 1:7]
##    d0101 d0102 d0103 d0104 d0105 d0106 d0107
## 3     0     1     0     0     0     0     1
## 4     0     0     0     1     0     0     1
```

To reduce the size of the dataset, we will aggregate by summing over five consecutive days. We would model the original dataset with a Bernoulli, therefore the aggregated dataset is modeled by a binomial (because a sum of Bernoulli variables is binomial distributed). There will be many 5 day blocks with less than 5 observations, because of missing values, and these will give binomials with less than 5 trials.

First, a new index is created to group the days in groups of five days:

```
id5 = rep(1:(365 / 5), each = 5)
```

The number of raining days is obtained with:

```
y5 <- t(apply(PRoccurrence[, 1:365], 1, tapply, id5, sum,
  na.rm = TRUE))
table(y5)
## y5
##     0     1     2     3     4     5
## 17227 10608  7972  4539  3063  1559
```

Next, the number of days with observed data in each group of 5 days, i.e. the trials in our binomial likelihood, is computed as:

```
n5 <- t(apply(!is.na(PRprec[, 3 + 1:365]), 1, tapply, id5, sum))
table(as.vector(n5))
##
##     0     1     2     3     4     5
##  3563    77    72    95   172 40989
```

Now, the aggregated data has 73 time points.

From the table above, it can be seen how there are 3563 periods of five days with no data recorded. The first approach when dealing with these missing values could be to remove such pairs of data, both y and n. If these are not removed, value NA has to be assigned to y when $n = 0$. However, n needs to be assigned a positive value (e.g., five). This is done as follows:

```
y5[n5 == 0] <- NA
n5[n5 == 0] <- 5
```

We set up all the variables in a dataframe:

```
n <- nrow(PRprec)
df = data.frame(y = as.vector(y5), ntrials = as.vector(n5),
  locx = rep(PRprec[, 1], k),
  locy = rep(PRprec[, 2], k),
  time = rep(1:k, each = n),
  station.id = rep(1:n, k))

summary(df)
##        y             ntrials          locx            locy
##  Min.   :0      Min.   :1.00    Min.   :-54.5    Min.   :-26.9
##  1st Qu.:0      1st Qu.:5.00    1st Qu.:-52.9    1st Qu.:-25.6
##  Median :1      Median :5.00    Median :-51.7    Median :-24.9
```

```
## Mean    :1        Mean    :4.98    Mean     :-51.6    Mean     :-24.7
## 3rd Qu.:2        3rd Qu.:5.00    3rd Qu. :-50.4    3rd Qu.:-23.8
## Max.    :5        Max.    :5.00    Max.     :-48.2    Max.     :-22.5
## NA's    :3563
##           time        station.id
## Min.    : 1    Min.    :  1
## 1st Qu.:19    1st Qu.:155
## Median :37    Median :308
## Mean    :37    Mean    :308
## 3rd Qu.:55    3rd Qu.:462
## Max.    :73    Max.    :616
##
```

7.3.2   Reducing the temporal resolution

This approach can be seen in the template code in Section 3.2 in Lindgren and
Rue (2015) and has also been considered in the last example in Blangiardo and
Cameletti (2015). The main idea is to place some knots over the time window
and define the model at such knots. Then, the projection is defined from the
time knots, similarly as is done for the spatial case with the mesh.

The knots are placed at every 6 time points of the temporally aggregated data,
which has 73 time points altogether. So, in the end there are only 12 knots
over time.

```
bt <- 6
gtime <- seq(1 + bt, k, length = round(k / bt)) - bt / 2
mesh.t <- inla.mesh.1d(gtime, degree = 1)
```

The model dimension is then 1152.

Then, when the projection matrix is computed, it is necessary to consider the
temporal mesh and the group index in the scale of the data to be analyzed.
The projection matrix with the definition of the spatial and temporal meshes
used above can be obtained as follows:

```
Ast <- inla.spde.make.A(mesh = mesh.s,
  loc = cbind(df$locx, df$locy), group.mesh = mesh.t,
  group = df$time)
```

The index set and the data stack are built as usual:

```
idx.st <- inla.spde.make.index('i', n.spde = spde.s$n.spde,
  n.group = mesh.t$n)
stk <- inla.stack(
  data = list(y = df$y, ntrials = df$ntrials),
  A = list(Ast, 1),
  effects = list(idx.st, data.frame(mu0 = 1,
    altitude = rep(PRprec$Altitude / 1000, k))))
```

Note that in the previous code the values of the altitude have been rescaled by dividing them by 1000. In general, we need to rescale the covariates to get stable numerical inference.

The formula is also the usual for a separable spatio-temporal model:

```
form <- y ~ 0 + mu0 + altitude + f(i, model = spde.s,
  group = i.group, control.group = list(model = 'ar1',
    hyper=list(theta=list(prior='pc.cor1', param=c(0.7, 0.7)))))
```

In order to reduce computational time in this example, a number of options will be set in the call to `inla()`. In particular, the adaptive Gaussian approximation (`strategy = 'adaptive'`) and the Empirical Bayes integration strategy over the hyperparameters (`int.strategy = 'eb'`) will be used. These options are passed in the `control.inla` argument of `inla()`. Furthermore, we start the optimizer at the initial values `init`:

```
# Initial values of hyperparameters
init = c(-0.5, -0.9, 2.6)
# Model fitting
result <- inla(form, 'binomial', data = inla.stack.data(stk),
  Ntrials = inla.stack.data(stk)$ntrials,
  control.predictor = list(A = inla.stack.A(stk), link = 1),
  control.mode = list(theta = init, restart=TRUE),
  control.inla = list(strategy = 'adaptive', int.strategy = 'eb'))
```

The fitted spatial effect can be plotted for each temporal knot and overlay the proportion of raining days considering the data closest to the time knots. First, a grid to make the projection is required:

```
data(PRborder)
r0 <- diff(range(PRborder[, 1])) / diff(range(PRborder[, 2]))
prj <- inla.mesh.projector(mesh.s, xlim = range(PRborder[, 1]),
  ylim = range(PRborder[, 2]), dims = c(100 * r0, 100))
in.pr <- inout(prj$lattice$loc, PRborder)
```

Next, the projection of the posterior mean fitted at each time knot is computed:

```
mu.st <- lapply(1:mesh.t$n, function(j) {
  idx <- 1:spde.s$n.spde + (j - 1) * spde.s$n.spde
  r <- inla.mesh.project(prj,
    field = result$summary.ran$i$mean[idx])
  r[!in.pr] <- NA
  return(r)
})
```

These projections are shown in Figure 7.6. We add the locations points with point size proportional to the rain occurrence in each period.

7.4   Conditional simulation: Combining two meshes

7.4.1   Motivation

There are a number of prediction problems that require modeling and prediction of the spatial, or spatio-temporal phenomenon over an extensive region, such as a country. However, in many situations, observations are only available in a limited part of the region. In this section we will discuss how to deal with this in a computationally efficient manner. Although practical computational issues are more common when dealing with spatio-temporal data, the example developed here focuses on spatial data to simplify the presentation. The same principle can be, however, easily extended to the spatio-temporal case.

We illustrate the case where there is a process that is only observed in part of the study region by using locations from Paraná state in Brazil. We use the boundary domain of the Paraná state and assume that we only have data from the left half part of the Paraná, while spatial prediction is desired for the entire state. Additionally, we consider fitting the model using a mesh around the available data and predicting using a mesh over the entire area of interest. These has been obtained with the following code and the resulting dataset has been plotted in Figure 7.7.

```
# loading the Parana border and precipitation data
data(PRborder)
data(PRprec)

# Which data points fall inside the left 50% of Parana
mid.long <- mean(range(PRborder[, 1]))
```

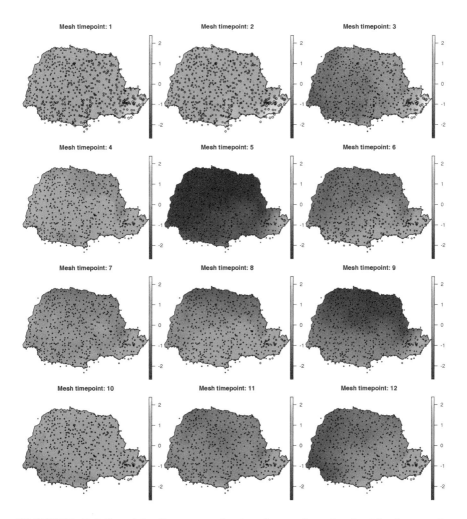

FIGURE 7.6 Spatial effect at each time knot obtanined with the spatio-temporal model fitted to the number of raining days in Paraná state (Brazil).

```
sel.loc <- which(PRprec[, 1] < mid.long)
sel.bor <- which(PRborder[, 1] < mid.long)
```

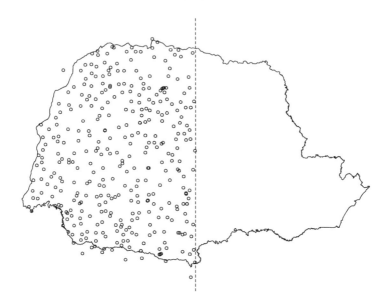

FIGURE 7.7 Problem setting: available data over half of the domain.

One way to obtain such predictions is to fit the model using a mesh that contains both, the locations where data is observed and the prediction locations (denoted by mesh2). Here we show a more efficient way, where the model is fitted using a mesh only around where the observations are placed (denoted by mesh1), and then conditional simulations are used to predict at the nodes of mesh2. This is achieved by taking advantage of numerical methods for sparse matrices applied to conditional simulation of GRMFs and the result is a considerable speedup in the computations when comparing to fitting the model directly using mesh2. After conditioning, the predictions at the data locations will be exactly the same values obtained from the fit using mesh1. In the geostatistics literature, this can be achieved with conditioning by kriging. The basis of the conditioning approach is to use the same covariance function for both the unconditional simulations and the predictions.

This is equivalent to the problem of sampling from a GMRF under the linear constraint

$$\mathbf{Ax} = \mathbf{b}, \tag{7.1}$$

where $\mathbf{A}$ is a $n_1 \times n_2$ matrix, with n_1 and n_2 being the number of nodes in mesh1 and mesh2, respectively. The vector $\mathbf{b}$ is the vector of constraints of length n_1 and corresponds to the predicted latent field from the fit using `mesh1`.

A way to obtain the correct conditional distribution of $\mathbf{x}^*$ is to sample from the unconstrained GMRF $\mathbf{x} \sim N(\mu, \mathbf{Q}^{-1})$ and then compute

$$\mathbf{x}^* = \mathbf{x} - \mathbf{Q}^{-1}\mathbf{A}^T(\mathbf{A}\mathbf{Q}^{-1}\mathbf{A}^T)^{-1}(\mathbf{A}\mathbf{x} - \mathbf{b}), \qquad (7.2)$$

where $\mathbf{Q}$ is a $n_2 \times n_2$ precision matrix and $\mathbf{A}\mathbf{Q}^{-1}\mathbf{A}^T$ is a dense matrix with dimensions equal to the number of constrains, that is, $n_1 \times n_1$. To factorize $\mathbf{A}\mathbf{Q}^{-1}\mathbf{A}^T$, we can take advantage of the sparse structure of $\mathbf{Q}$ and obtain fast computations for $n_1 \ll n_2$.

7.4.2 Paraná state example

At each location i, we suppose the following distribution for the data y_i:

$$y_i \sim N(\eta_i, \sigma_\epsilon), \qquad (7.3)$$

with σ_ϵ being an iid Gaussian noise and η_i being the linear predictor, defined as:

$$\eta_i = \beta_0 + u_i, \qquad (7.4)$$

where β_0 is the intercept and u_i is a realization of a spatial random Gaussian field with Matérn covariance at the data locations i.

We use simulated data to show how to obtain predictions in `mesh2` after fitting the model using `mesh1`. To illustrate the ability of our method to predict using a different mesh, we assume that the data comes from the random field based on `mesh2`, namely u_i.

We build `mesh1` considering only the data from the western half of Paraná state:

```
#  Building mesh1
mesh1 <- inla.mesh.2d(loc = PRprec[sel.loc, 1:2], max.edge = 1,
  cutoff = 0.1, offset = 1.2)
```

Next, we create `mesh2` such that all the nodes from `mesh1` are also nodes of `mesh2`. In addition, we use the border of the Paraná state to define the high resolution interior of the `mesh2`. In order to implement both these restrictions we first create an auxiliary mesh `mesh2a` considering the border of the Paraná state. Then we use the locations from `mesh1` and from the auxiliary mesh to create `mesh2`.

```
# define a boundary for the Parana border and an auxiliary mesh
ibound <- inla.nonconvex.hull(PRborder, 0.05, 2, resol = 250)
mesh2a <- inla.mesh.2d(mesh1$loc, boundary = ibound,
  max.edge = 0.2, cutoff = 0.1)
# Building mesh2 considering wider a boundary
bound <- inla.nonconvex.hull(PRborder, 2)
mesh2 <- inla.mesh.2d(loc = rbind(mesh1$loc, mesh2a$loc),
  boundary = bound, max.edge = 1, cutoff = 0.1)
```

As a result, mesh1 has 379 nodes and mesh2 has 1477 nodes. We can see these
two meshes in Figure 7.8.

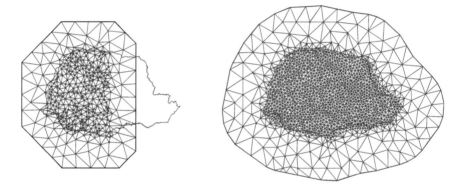

FIGURE 7.8 The first mesh, together with the data locations in blue (left).
Mesh mesh2, which will be used for predictions, and the points of the first mesh
represented as red points (right). The inner blue polygon shows the Paraná
state border.

To simulate data, we need to fix the range and standard deviation and then
define the SPDE models for both meshs as follows:

```
# SPDE model parameters
range <- 3
std.u <- 1
# Define the SPDE models for mesh1 and mesh2
spde1 = inla.spde2.pcmatern(mesh1, prior.range = c(1, 0.1),
  prior.sigma = c(1, 0.1))
spde2 = inla.spde2.pcmatern(mesh2, prior.range = c(1, 0.1),
  prior.sigma = c(1, 0.1))
```

The precision matrices for both SPDE models are built with:

```
# Obtain the precision matrix for spde1 and spde2
Q1 = inla.spde2.precision(spde1,
  theta = c(log(range), log(std.u)))
Q2 = inla.spde2.precision(spde2,
  theta = c(log(range), log(std.u)))
```

Simulation of the random field at the nodes of `mesh2` can be performed as follows:

```
u <- as.vector(inla.qsample(n = 1, Q = Q2, seed = 1))
```

We complete the data simulation by projecting the mesh nodes into the observed data points and adding an iid noise. We also build the projection matrix for `mesh1`, which will be used for fitting the model:

```
A1 <- inla.spde.make.A(mesh1,
  loc = as.matrix(PRprec[sel.loc, 1:2]))
A2 <- inla.spde.make.A(mesh2,
  loc = as.matrix(PRprec[sel.loc, 1:2]))
```

We now sample the spatial field and iid Gaussian noise at the observation locations:

```
std.epsilon = 0.1
y <- drop(A2 %*% u) + rnorm(nrow(A2), sd = std.epsilon)
```

The stack data includes the intercept and the SPDE model defined at `mesh1`:

```
stk <- inla.stack(
  data = list(resp = y),
  A = list(A1, 1),
  effects = list(i = 1:spde1$n.spde,m = rep(1, length(y))),
  tag = 'est')
```

7.4.3 Fitting the model

The model is fitted as follows:

```
res <- inla(resp ~ 0 + m + f(i, model = spde1),
  data = inla.stack.data(stk),
```

```
control.compute = list(config = TRUE),
control.predictor = list(A = inla.stack.A(stk)))
```

The summary of the posterior marginal distribution for the standard deviation
of the noise, the range and the standard deviation, together with the true
values is shown below.

```
# Marginal for standard deviation of Gaussian likelihood
p.s.eps <- inla.tmarginal(function(x) 1 / sqrt(exp(x)),
  res$internal.marginals.hyperpar[[1]])
# Summary of post. marg. of st. dev.
s.std <- unlist(inla.zmarginal(p.s.eps, silent = TRUE))[c(1:3, 7)]

hy <- cbind(True = c(std.epsilon, range, std.u),
  rbind(s.std, res$summary.hyperpar[2:3, c(1:3, 5)]))
rownames(hy) <- c('Std epsilon', 'Range field', 'Std field')
hy
##                True    mean        sd 0.025quant 0.975quant
## Std epsilon     0.1  0.1085  0.007258    0.09497     0.1235
## Range field     3.0  2.3078  0.542444    1.50645     3.6146
## Std field       1.0  0.7203  0.146294    0.49784     1.0673
```

7.4.4   Obtaining predictions

Before proceeding to the actual prediction, we need to sample from the posterior
distribution using the fitted model. We draw 100 samples from the posterior
considering the internal parametrization for the hyperparameters with:

```
nn <- 100
s <- inla.posterior.sample(n = nn, res, intern = TRUE,
  seed = 1, add.names = FALSE)
```

We can find the indices for the spatial random effect i in the following way:

```
## Find the values of latent field "i" in samples from mesh1
contents <- res$misc$configs$contents
effect <- "i"
id.effect <- which(contents$tag == effect)
ind.effect <- contents$start[id.effect] - 1 +
  (1:contents$length[id.effect])
```

For each sample from the posterior distribution, the following code produces predictions of the latent field **u** at the nodes of `mesh2` constrained on the predictions at the nodes of `mesh1` being equal of the values of the latent field from the posterior samples generated from fitting the model with `mesh1`. This code is based on Equation (7.2), but with additional complications to achieve greater computational speed:

```
# Obtain predictions at the nodes of mesh2
loc1 = mesh1$loc[,1:2]
loc2 = mesh2$loc[,1:2]
n = mesh2$n

mtch = match(data.frame(t(loc2)), data.frame(t(loc1)))
idx.c = which(!is.na(mtch))
idx.u = setdiff(1:mesh2$n, idx.c)
p = c(idx.u, idx.c)

ypred.mesh2 = matrix(c(NA), mesh2$n, nn)

m <- n - length(idx.c)
iperm <- numeric(m)

t0 <- Sys.time()
for(ind in 1:nn){

  Q.tmp = inla.spde2.precision(spde2,
    theta = s[[ind]]$hyperpar[2:3])

  Q = Q.tmp[p, p]
  Q.AA = Q[1:m, 1:m]
  Q.BB = Q[(m + 1):n, (m + 1):n]
  Q.AB = t(Q[(m + 1):n, 1:m])
  Q.AA.sf = Cholesky(Q.AA, perm = TRUE, LDL = FALSE)
  perm = Q.AA.sf@perm + 1
  iperm[perm] = 1:m
  x = solve(Q.AA.sf, rnorm(m), system = "Lt")
  xc = s[[ind]]$latent[ind.effect]
  xx = solve(Q.AA.sf, -Q.AB %*% xc, system = "A")

  x = rep(NA, n)
  x[idx.u] = c(as.matrix(xx))
  x[idx.c] = xc

  ypred.mesh2[, ind] = x
}
```

```
Sys.time() - t0
## Time difference of 3.629 secs
```

Notice that the code above computes the Cholesky factorization of the precision matrix **Q** rapidly by taking the sparsity of this matrix into account. This makes it possible to obtain fast predictions in a large number of locations. Another possibility is to fit the model directly using `mesh2` instead of `mesh1`. However, this would require a lot more computational time and results would be similar to the procedure shown here.

To produce maps of the predicted random field in a fine grid, we compute a projection matrix for a grid of points over a square that contains the locations of the border of the Paraná state.

```
# Projection from the mesh nodes to a fine grid
projgrid  <- inla.mesh.projector(mesh2,
  xlim = range(PRborder[, 1]), ylim = range(PRborder[, 2]),
  dims = c(300, 300))
```

Then, we can obtain the projection of the simulated random field and compare with the projection of the posterior mean of the predictions. Missing values are assigned to the grid points that are outside of the Paraná state:

```
# Find points inside the state of Parana
xy.in <- inout(projgrid$lattice$loc, PRborder[, 1:2])
# True field
r1 <- inla.mesh.project(projgrid , field = u)
r1[!xy.in] <- NA

# Mean predicted random field
r2 <- inla.mesh.project(projgrid , field = rowMeans(ypred.mesh2))
r2[!xy.in] <- NA

# sd of the predicted random field
sd.r2 <- inla.mesh.project(
  projgrid, field=apply(ypred.mesh2, 1, sd, na.rm = TRUE))
sd.r2[!xy.in] <- NA

# plotting
par(mfrow = c(3, 1), mar = c(0, 0, 0, 0))
zlm <- range(c(r1, r2), na.rm = TRUE)

# Map of the true field
```

```
book.plot.field(list(x = projgrid$x, y = projgrid$y, z = r1),
  zlim = zlm)
points(PRprec[sel.loc, 1:2], col = "black", asp = 1, cex = 0.3)

## Map of the mean of the mean predicted random field
book.plot.field(list(x = projgrid$x, y = projgrid$y, z = r2),
  zlim = zlm)
points(PRprec[sel.loc, 1:2], col = "black", asp = 1, cex = 0.3)

book.plot.field(list(x = projgrid$x, y = projgrid$y, z = sd.r2))
points(PRprec[sel.loc, 1:2], col = "black", asp = 1, cex = 0.3)
```

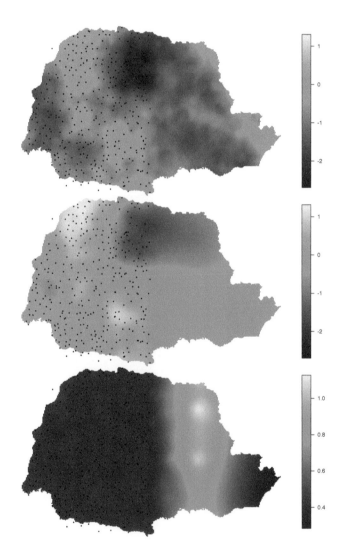

FIGURE 7.9 The simulated field (top), the estimated posterior mean (middle) and the posterior marginal standard deviation (bottom).

8

Space-time applications

In this chapter we generalize some of the examples presented in the book so far to space-time. In particular, we consider space-time coregionalization models, dynamic regression models, space-time point processes, and space-time Hurdle models.

8.1 Space-time coregionalization model

In this section we generalize the coregionalization model found in Section 3.1 to a space-time model. This model is very similar, but much more computationally demanding. Because of this, we use a cruder mesh in our example than what we would usually recommend.

8.1.1 Model and parametrization

The model is similar to the spatial model:

$$y_1(s,t) = \alpha_1 + z_1(s,t) + e_1(s,t)$$

$$y_2(s,t) = \alpha_2 + \lambda_1 z_1(s,t) + z_2(s,t) + e_2(s,t)$$

$$y_3(s,t) = \alpha_3 + \lambda_2 z_1(s,t) + \lambda_3 z_2(s,t) + z_3(s,t) + e_3(s,t)$$

Here, $z_k(s,t)$ are space-time effects and $e_k(s,t)$ are uncorrelated error terms, $k = 1, 2, 3$.

8.1.2 Data simulation

First of all, these are the values of the parameters that will be used to simulate the data:

```
alpha <- c(-5, 3, 10) # intercept on reparametrized model
z.sigma = c(0.5, 0.6, 0.7) # random field marginal std
range = c(0.2, 0.3, 0.4) # GRF scales: range parameters
beta <- c(0.7, 0.5, -0.5) # copy par.: reparam. coreg. par.
rho <- c(0.7, 0.8, 0.9) # temporal correlations
n <- 50 # number of spatial locations
k <- 4  # number of time points
e.sigma <- c(0.3, 0.2, 0.15) # The measurement error marginal std
```

We use the same spatial and temporal locations for all response variables. Note that in Section 3.1 we use different spatial locations, and that it is also possible to use different time points (when using a temporal mesh).

The locations are simulated as follows:

```
loc <- cbind(runif(n), runif(n))
```

Then, the book.rMatern() function defined in Section 2.1.4 will be used to simulate independent random field realizations for each time:

```
x1 <- book.rMatern(k, loc, range = range[1], sigma = z.sigma[1])
x2 <- book.rMatern(k, loc, range = range[2], sigma = z.sigma[2])
x3 <- book.rMatern(k, loc, range = range[3], sigma = z.sigma[3])
```

The temporal dependency is modeled as an autoregressive first order process, the same as was used in Chapter 7.

```
z1 <- x1
z2 <- x2
z3 <- x3

for (j in 2:k) {
    z1[, j] <- rho[1] * z1[, j - 1] + sqrt(1 - rho[1]^2) * x1[, j]
    z2[, j] <- rho[2] * z2[, j - 1] + sqrt(1 - rho[2]^2) * x2[, j]
    z3[, j] <- rho[3] * z3[, j - 1] + sqrt(1 - rho[3]^2) * x3[, j]
}
```

We use the constants $\sqrt{(1 - \rho_j^2)}$, $j = 1, 2, 3$ to ensure that the samples are taken from the stationary distribution.

Then the response variables are sampled:

```
y1 <- alpha[1] + z1 + rnorm(n, 0, e.sigma[1])
y2 <- alpha[2] + beta[1] * z1 + z2 + rnorm(n, 0, e.sigma[2])
y3 <- alpha[3] + beta[2] * z1 + beta[3] * z2 + z3 +
  rnorm(n, 0, e.sigma[3])
```

8.1.3 Model fitting

We define a crude mesh to save computational time:

```
mesh <- inla.mesh.2d(loc, max.edge = 0.2, offset = 0.1,
  cutoff = 0.1)
```

Similarly as in previous examples, the SPDE model will consider the PC-priors derived in Fuglstad et al. (2018) for the model parameters as the range, $\sqrt{8\nu}/\kappa$, and the marginal standard deviation. These are set when defining the SPDE latent effect:

```
spde <- inla.spde2.pcmatern(mesh = mesh,
  prior.range = c(0.05, 0.01), # P(range < 0.05) = 0.01
  prior.sigma = c(1, 0.01)) # P(sigma > 1) = 0.01
```

Indices for the space-time fields and for the copies need to be defined as well. As the same mesh is considered in all effects, these indices are the same for all the effects:

```
s1 <- rep(1:spde$n.spde, times = k)
s2 <- s1
s3 <- s1
s12 <- s1
s13 <- s1
s23 <- s1

g1 <- rep(1:k, each = spde$n.spde)
g2 <- g1
g3 <- g1
g12 <- g1
g13 <- g1
g23 <- g1
```

The prior on ρ_j is chosen as a Penalized Complexity prior (Simpson et al., 2017) as well:

```
rho1p <- list(theta = list(prior = 'pccor1', param = c(0, 0.9)))
ctr.g <- list(model = 'ar1', hyper = rho1p)
```

The prior above is chosen to consider $P(\rho_j > 0) = 0.9$.

Priors for each of the copy parameters are Gaussian with zero mean and precision 10:

```
hc1 <- list(theta = list(prior = 'normal', param = c(0, 10)))
```

The formula, which includes all the terms in the model and the priors previously defined, is:

```
form <- y ~ 0 + intercept1 + intercept2 + intercept3 +
  f(s1, model = spde, group = g1, control.group = ctr.g) +
  f(s2, model = spde, group = g2, control.group = ctr.g) +
  f(s3, model = spde, group = g3, control.group = ctr.g) +
  f(s12, copy = "s1", group = g12, fixed = FALSE, hyper = hc1) +
  f(s13, copy = "s1", group = g13, fixed = FALSE, hyper = hc1) +
  f(s23, copy = "s2", group = g23, fixed = FALSE, hyper = hc1)
```

The projector matrix is defined as:

```
stloc <- kronecker(matrix(1, k, 1), loc) # repeat coord. each time
A <- inla.spde.make.A(mesh, stloc, n.group = k,
  group = rep(1:k, each = n))
```

Note that in this example the projector matrices (the **A**-matrix) are all equal for the different time points because all points have the same coordinates at different times, but the projector matrix can be different when observations at different times are at different locations.

Then data are organized in three data stacks, which are joined:

```
stack1 <- inla.stack(
  data = list(y = cbind(as.vector(y1), NA, NA)),
  A = list(A),
  effects = list(list(intercept1 = 1, s1 = s1, g1 = g1)))

stack2 <- inla.stack(
  data = list(y = cbind(NA, as.vector(y2), NA)),
  A = list(A),
```

```
effects = list(list(intercept2 = 1, s2 = s2, g2 = g2,
  s12 = s12, g12 = g12)))

stack3 <- inla.stack(
  data = list(y = cbind(NA, NA, as.vector(y3))),
  A = list(A),
  effects = list(list(intercept3 = 1, s3 = s3, g3 = g3,
    s13 = s13, g13 = g13, s23 = s23, g23 = g23)))

stack <- inla.stack(stack1, stack2, stack3)
```

Another PC-prior is considered for the precision of the errors (Simpson et al., 2017) in the three likelihoods in the model:

```
eprec <- list(hyper = list(theta = list(prior = 'pc.prec',
  param = c(1, 0.01))))
```

This model has 15 hyperparameters. To make the optimization process fast, the parameter values used in the simulation will be used as the initial values:

```
theta.ini <- c(log(1 / e.sigma^2),
  c(log(range), log(z.sigma),
  qlogis(rho))[c(1, 4, 7, 2, 5, 8, 3, 6, 9)], beta)

# We jitter the starting values to avoid artificially
# recovering the true values
theta.ini = theta.ini + rnorm(length(theta.ini), 0, 0.1)
```

Then, the model is fitted with:

```
result <- inla(form, rep('gaussian', 3),
  data = inla.stack.data(stack),
  control.family = list(eprec, eprec, eprec),
  control.mode = list(theta = theta.ini, restart = TRUE),
  control.inla = list(int.strategy = 'eb'),
  control.predictor = list(A = inla.stack.A(stack)))
```

Computation time for this model in seconds is:

```
##     Pre Running    Post   Total
##   5.495  47.866   0.236  53.597
```

Table 8.1 summarizes the posterior marginal distributions of the parameters in the model. These include the intercepts, precisions of the errors, temporal correlations, copy parameters, and range and standard deviations of the random fields.

TABLE 8.1: Summary of the posterior distributions of the parameters in the model.

Parameter	True	Mean	St. Dev.	2.5% quant.	97.5% quant.
intercept1	-5.00	-5.0261	0.1344	-5.2901	-4.7624
intercept2	3.00	3.1673	0.2071	2.7607	3.5736
intercept3	10.00	9.7655	0.2802	9.2154	10.3152
e1	11.11	16.0295	2.3083	11.9426	21.0135
e2	25.00	14.6702	2.4464	10.4295	20.0336
e3	44.44	15.2115	2.2840	11.1578	20.1285
GroupRho for s1	0.70	0.8737	0.0438	0.7707	0.9411
GroupRho for s2	0.80	0.9040	0.0355	0.8192	0.9569
GroupRho for s3	0.90	0.9829	0.0101	0.9577	0.9961
Beta for s12	0.70	0.6629	0.1294	0.4100	0.9184
Beta for s13	0.50	0.5157	0.1214	0.2790	0.7557
Beta for s23	-0.50	-0.5119	0.1357	-0.7823	-0.2479
Range for s1	0.20	0.1515	0.0401	0.0849	0.2416
Range for s2	0.30	0.2462	0.0622	0.1441	0.3872
Range for s3	0.40	0.2389	0.0612	0.1409	0.3802
Stdev for s1	0.50	0.7366	0.1246	0.5285	1.0161
Stdev for s2	0.60	0.6778	0.0918	0.5168	0.8767
Stdev for s3	0.70	0.9177	0.1224	0.7035	1.1832

The posterior mean for each random field is projected to the observation locations and shown against the simulated correspondent fields in Figure 8.1.

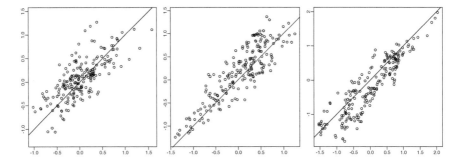

FIGURE 8.1 True and fitted random field values.

Remember that the crude mesh leads to a crude approximation for the spatial covariance. This is not recommended when fitting a model in practice. However, this setting can be considered to obtain initial results, and for illustrative code examples. In this particular case, it seems that the method provided reasonable estimates of the model parameters.

8.2 Dynamic regression example

There is large literature about dynamic models, which includes some books, such as West and Harrison (1997) and Petris et al. (2009). These models basically define a hierarchical framework for a class of time series models. A particular case is the dynamic regression model, where the regression coefficients are modeled as time series. That is the case when the regression coefficients vary smoothly over time.

8.2.1 Dynamic space-time regression

The specific class of models for spatially structured time series was proposed in Gelfand et al. (2003), where the regression coefficients vary smoothly over time and space. For the areal data case, the use of proper Gaussian Markov random fields (PGMRF) over space has been proposed in Vivar and Ferreira (2009). There exists a particular class of such models called "spatially varying coefficient models", in which the regression coefficients vary over space. See, for example, Assunção et al. (1999), Assunção et al. (2002) and Gamerman et al. (2003).

In Gelfand et al. (2003), the Gibbs sampler was used for inference and it was claimed that a better algorithm is needed due to strong autocorrelations. In Vivar and Ferreira (2009), the use of forward information filtering and backward sampling (FIFBS) recursions were proposed. Both MCMC algorithms are computationally expensive.

The FIFBS algorithm can be avoided as a relation between the Kalman-filter and the Cholesky factorization is proposed in Knorr-Held and Rue (2002). The Cholesky factorization is more general and has a better performance when using sparse matrix methods (p. 149, Rue and Held, 2005). Additionally, the restriction that the prior for the latent field has to be proper can be avoided.

When the likelihood is Gaussian, there is no approximation needed in the inference process since the distribution of the latent field given the data and the hyperparameters is Gaussian. So, the main task is to perform inference for the hyperparameters in the model. For this, the mode and curvature around can be found without any sampling method. For the class of models in Vivar and Ferreira (2009) it is natural to use `INLA`, as shown in Ruiz-Cárdenas et al.

(2012), and for the models in Gelfand et al. (2003), the SPDE approach can be used when considering the Matérn covariance for the spatial part.

In this example, it will be shown how to fit the space-time dynamic regression model as discussed in Gelfand et al. (2003), considering the Matérn spatial covariance and the AR(1) model for time, which corresponds to the exponential correlation function. This particular covariance choice corresponds to the model in Cameletti et al. (2013), where only the intercept is dynamic. Here, the considered case is that of a dynamic intercept and a dynamic regression coefficient for a harmonic over time.

8.2.2 Simulation from the model

In order to simulate some data to fit the model, the spatial locations are sampled first, as follows:

```
n <- 150
set.seed(1)
coo <- matrix(runif(2 * n), n)
```

To sample from a random field at a set of locations, the book.rMatern() function defined in the Section 2.1.4 will be used to simulate independent random field realizations for each time.

k (number of time points) samples will be drawn from the random field. Then, they are temporally correlated considering the time autoregression:

```
kappa <- c(10, 12)
sigma2 <- c(1 / 2, 1 / 4)
k <- 15
rho <- c(0.7, 0.5)

set.seed(2)
beta0 <- book.rMatern(k, coo, range = sqrt(8) / kappa[1],
  sigma = sqrt(sigma2[1]))

set.seed(3)
beta1 <- book.rMatern(k, coo, range = sqrt(8) / kappa[2],
  sigma = sqrt(sigma2[2]))
beta0[, 1] <- beta0[, 1] / (1 - rho[1]^2)
beta1[, 1] <- beta1[, 1] / (1 - rho[2]^2)

for (j in 2:k) {
  beta0[, j] <- beta0[, j - 1] * rho[1] + beta0[, j] *
    (1 - rho[1]^2)
```

```
    beta1[, j] <- beta1[, j - 1] * rho[2] + beta1[, j] *
      (1 - rho[2]^2)
}
```

Here, the $(1 - \rho_j^2)$ term appears because it is in parametrization of the AR(1) model in INLA.

To get the response, the harmonic is defined as a function over time, and then the mean and the error terms are added up:

```
set.seed(4)
# Simulate the covariate values
hh <- runif(n * k)
mu.beta <- c(-5, 1)
taue <- 20

set.seed(5)
# Error in the observation
error <- rnorm(n * k, 0, sqrt(1 / taue))
# Dynamic regression part
y <- (mu.beta[1] + beta0) + (mu.beta[2] + beta1) * hh +
  error
```

8.2.3 Fitting the model

There are two space-time terms in the model, each one with three hyperparameters: precision, spatial scale and temporal scale (or temporal correlation). So, considering the likelihood precision, there are 7 hyperparameters in total. To perform fast inference, a crude mesh with a small number of vertices is chosen:

```
mesh <- inla.mesh.2d(coo, max.edge = c(0.15, 0.3),
  offset = c(0.05, 0.3), cutoff = 0.07)
```

This mesh has 195 points.

As in previous examples, the SPDE model will consider the PC-priors derived in Fuglstad et al. (2018) for the model parameters as the practical range, $\sqrt{8\nu}/\kappa$, and the marginal standard deviation:

```
spde <- inla.spde2.pcmatern(mesh = mesh,
  prior.range = c(0.05, 0.01), # P(practic.range < 0.05) = 0.01
  prior.sigma = c(1, 0.01)) # P(sigma > 1) = 0.01
```

A different index is needed for each call to the f() function, even if they are the same, so:

```
i0 <- inla.spde.make.index('i0', spde$n.spde, n.group = k)
i1 <- inla.spde.make.index('i1', spde$n.spde, n.group = k)
```

In the SPDE approach, the space-time model is defined at a set of mesh nodes. As a continuous time is being considered, it is also defined on a set of time knots. So, it is necessary to deal with the projection from the model domain (nodes, knots) to the space-time data locations. For the intercept, it is the same way as in previous examples. For the regression coefficients, all that is required is to multiply the projector matrix by the covariate vector column, i. e., each column of the projector matrix is multiplied by the covariate vector. This can be seen from the following structure of the linear predictor η:

$$
\begin{aligned}
\eta &= \mu_{\beta_0} + \mu_{\beta_2} \boldsymbol{h} + \boldsymbol{A}\boldsymbol{\beta}_0 + (\boldsymbol{A}\boldsymbol{\beta}_1)\boldsymbol{h} \\
&= \mu_{\beta_0} + \mu_{\beta_1} \boldsymbol{h} + \boldsymbol{A}\boldsymbol{\beta}_0 + (\boldsymbol{A} \oplus (\boldsymbol{h}\mathbf{1}^\top))\boldsymbol{\beta}_1
\end{aligned}
$$

Here, $\boldsymbol{A} \oplus (\boldsymbol{h}\mathbf{1}^\top)$ is the row-wise Kronecker product between $\boldsymbol{A}$ and vector $\boldsymbol{h}$ (with length equal the number of rows in $\boldsymbol{A}$) expressed as the Kronecker sum of $\boldsymbol{A}$ and $\boldsymbol{h}\mathbf{1}^\top$. This operation can be performed using the inla.row.kron() function and is done internally in the function inla.spde.make.A() when supplying a vector in the weights argument.

The space-time projector matrix $\boldsymbol{A}$ is defined as follows:

```
A0 <- inla.spde.make.A(mesh,
  cbind(rep(coo[, 1], k), rep(coo[, 2], k)),
  group = rep(1:k, each = n))
A1 <- inla.spde.make.A(mesh,
  cbind(rep(coo[, 1], k), rep(coo[, 2], k)),
  group = rep(1:k, each = n), weights = hh)
```

The data stack is as follows:

```
stk.y <- inla.stack(
  data = list(y = as.vector(y)),
  A = list(A0, A1, 1),
  effects = list(i0, i1, data.frame(mu1 = 1, h = hh)),
  tag = 'y')
```

Here, i0 is similar to i1 and variables mu1 and h in the second element of the effects data.frame are for μ_{β_0}, μ_{β_1} and μ_{β_2}.

The formula considered in this model takes the following effects into account:

```
form <- y ~ 0 + mu1 + h + # to fit mu_beta
  f(i0, model = spde, group = i0.group,
    control.group = list(model = 'ar1')) +
  f(i1, model = spde, group = i1.group,
    control.group = list(model = 'ar1'))
```

As the model considers a Gaussian likelihood, there is no approximation in the fitting process. The first step of the INLA algorithm is the optimization to find the mode of the 7 hyperparameters in the model. By choosing good starting values, fewer iterations will be needed in this optimization process. Below, starting values are defined for the hyperparameters in the internal scale considering the values used to simulate the data:

```
theta.ini <- c(
  log(taue), # likelihood log precision
  log(sqrt(8) / kappa[1]), # log range 1
  log(sqrt(sigma2[1])), # log stdev 1
  log((1 + rho[1])/(1 - rho[1])), # log trans. rho 1
  log(sqrt(8) / kappa[2]), # log range 1
  log(sqrt(sigma2[2])), # log stdev 1
  log((1 + rho[2]) / (1 - rho[2])))# log trans. rho 2
```

```
theta.ini
## [1]  2.9957 -1.2629 -0.3466  1.7346 -1.4452 -0.6931  1.0986
```

The integration step when using the CCD strategy will integrate over 79 hyperparameter configurations, as we have 7 hyperparameters. For complex models, model fitting may take a few minutes. A bigger `tolerance` value in `inla.control` can be set to reduce the number of posterior evaluations, which will also reduce computational time. However, in the following `inla()` call we avoid it by using an Empirical Bayes strategy.

Finally, model fitting considering the initial values defined above will be done as follows:

```
res <- inla(form, family = 'gaussian',
  data = inla.stack.data(stk.y),
  control.predictor = list(A = inla.stack.A(stk.y)),
  control.inla = list(int.strategy = 'eb'),# no integr. wrt theta
  control.mode = list(theta = theta.ini, # initial theta value
    restart = TRUE))
```

The time required to fit this model has been:

```
res$cpu
##          Pre   Running        Post     Total
##      3.0943 263.4097    0.4765 266.9805
```

Summary of the posterior marginals of μ_{β_1}, μ_{β_2} and the likelihood precision (i.e., $1/\sigma_e^2$) are available in Table 8.2.

TABLE 8.2: Summary of the posterior distributions of the parameters in the model.

Parameter	True	Mean	St. Dev.	2.5% quant.	97.5% quant.
μ_{β_1}	-5	-4.7789	0.2022	-5.1759	-4.382
μ_{β_2}	1	0.9303	0.0587	0.8151	1.046
$1/\sigma_e^2$	20	10.8340	0.4940	9.9005	11.841

The posterior marginal distributions for the range, standard deviation and autocorrelation parameter for each spatio-temporal process are in Figure 8.2.

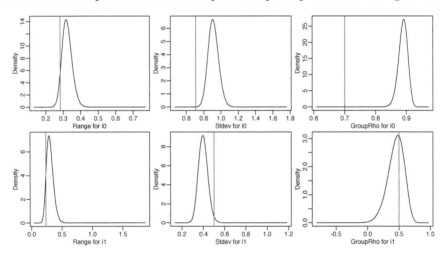

FIGURE 8.2 Posterior marginal distributions for the hyperparameters of the space-time fields. Red lines represent the true values of the parameters.

In order to look deeper into the posterior means of the dynamic coefficients, the correlation between the mean of the simulated values and the corresponding posterior means have been computed:

```
## Using A0 to account only for the coeff.
c(beta0 = cor(as.vector(beta0),
```

```
    drop(A0 %*% res$summary.ran$i0$mean)),
  beta1 = cor(as.vector(beta1),
    drop(A0 %*% res$summary.ran$i1$mean)))
## beta0 beta1
## 0.9434 0.6107
```

8.3 Space-time point process: Burkitt example

In this section a model for space-time point processes is developed and applied to a real dataset.

8.3.1 The dataset

The model developed in this section will be applied in the analysis of the `burkitt` dataset from the `splancs` package (Rowlingson and Diggle, 1993). This dataset records cases of Burkitt's lymphoma in the Western Nile district of Uganda during the period 1960-1975 (see, Bailey and Gatrell, 1995, Chapter 3). This dataset contains the five columns described in Table 8.3.

TABLE 8.3: Description of the `burkitt` dataset, which records cases of Burkitt's lymphoma in Uganda.

Variable	Description
x	Easting
y	Northing
t	Day, starting at 1/1/1960 of onset
age	age of child patient
dates	Day, as string yy-mm-dd

This dataset can be loaded as follows:

```
data('burkitt', package = 'splancs')
```

The spatial coordinates and time values can be summarized as follows:

```
t(sapply(burkitt[, 1:3], summary))
##    Min. 1st Qu. Median   Mean 3rd Qu. Max.
## x  255   269.0  282.5  286.3   300.2  335
```

```
## y  247   326.8  344.5  338.8   362.0  399
## t  413  2411.8 3704.5 3529.9  4700.2 5775
```

A set of knots over time needs to be defined in order to fit a SPDE spatio-temporal model. It is then used to build a temporal mesh, as follows:

```
k <- 6
tknots <- seq(min(burkitt$t), max(burkitt$t), length = k)
mesh.t <- inla.mesh.1d(tknots)
```

Figure 8.3 shows the temporal mesh as well as the times at which the events occurred.

FIGURE 8.3 Time when each event occurred (black) and knots used for inference (blue).

The spatial mesh can be created using the polygon of the region as a boundary. The domain polygon can be converted into a **SpatialPolygons** class with:

```
domainSP <- SpatialPolygons(list(Polygons(list(Polygon(burbdy)),
  '0')))
```

This boundary is then used to compute the mesh:

```
mesh.s <- inla.mesh.2d(burpts,
  boundary = inla.sp2segment(domainSP),
  max.edge = c(10, 25), cutoff = 5) # a crude mesh
```

Again, the SPDE model is defined to use the PC-priors derived in Fuglstad et al. (2018) for the range and the marginal standard deviation. These are defined now:

```
spde <- inla.spde2.pcmatern(mesh = mesh.s,
  prior.range = c(5, 0.01), # P(practic.range < 5) = 0.01
  prior.sigma = c(1, 0.01)) # P(sigma > 1) = 0.01
m <- spde$n.spde
```

The spatio temporal projection matrix is made considering both spatial and temporal locations and both spatial and temporal meshes, as follows:

```
Ast <- inla.spde.make.A(mesh = mesh.s, loc = burpts,
  n.group = length(mesh.t$n), group = burkitt$t,
  group.mesh = mesh.t)
```

The dimension of the resulting projector matrix is:

```
dim(Ast)
## [1]   188 2424
```

Internally, the `inla.spde.make.A()` function makes a row Kronecker product (see manual page of function `inla.row.kron()`) between the spatial projector matrix and the group (temporal dimension, in our case) projector one. This matrix has number of columns equal to the number of nodes in the mesh times the number of groups.

The index set is made considering the group feature:

```
idx <- inla.spde.make.index('s', spde$n.spde, n.group = mesh.t$n)
```

The data stack can be made considering the ideas for the purely spatial model. So, it is necessary to consider the expected number of cases at the integration points and the data locations. For the integration points, it is the space-time volume computed for each mesh node and time knot, considering the spatial area of the dual mesh polygons, as in Chapter 4, times the length of the time window at each time point. For the data locations, it is zero as for a point the expectation is zero, as in the likelihood approximation proposed by Simpson et al. (2016).

The dual mesh is extracted considering function `book.mesh.dual()`, available in file **spde-book-functions.R**, as follows:

```
dmesh <- book.mesh.dual(mesh.s)
```

Then, the intersection with each polygon from the dual mesh is computed using functions `gIntersection()`, from the **rgeos** package, as:

```
library(rgeos)
w <- sapply(1:length(dmesh), function(i) {
  if (gIntersects(dmesh[i,], domainSP))
    return(gArea(gIntersection(dmesh[i,], domainSP)))
```

```
  else return(0)
})
```

The sum of all the weights is equal to 1.1035×10^4. This is the same as the domain area:

```
gArea(domainSP)
## [1] 11035
```

The spatio-temporal volume is the product of these values and the time window length of each time knot. It is computed here:

```
st.vol <- rep(w, k) * rep(diag(inla.mesh.fem(mesh.t)$c0), m)
```

The data stack is built using the following lines of R code:

```
y <- rep(0:1, c(k * m, n))
expected <- c(st.vol, rep(0, n))
stk <- inla.stack(
  data = list(y = y, expect = expected),
  A = list(rbind(Diagonal(n = k * m), Ast), 1),
  effects = list(idx, list(a0 = rep(1, k * m + n))))
```

Finally, model fitting will be done using the cruder Gaussian approximation:

```
pcrho <- list(prior = 'pccor1', param = c(0.7, 0.7))
form <- y ~ 0 + a0 + f(s, model = spde, group = s.group,
  control.group = list(model = 'ar1',
    hyper = list(theta = pcrho)))

burk.res <- inla(form, family = 'poisson',
  data = inla.stack.data(stk), E = expect,
  control.predictor = list(A = inla.stack.A(stk)),
  control.inla = list(strategy = 'adaptive'))
```

The exponential of the intercept plus the random effect at each space-time integration point is the relative risk at each of these points. This relative risk times the space-time volume will give the expected number of points $(E(n))$ at each one of these space-time locations. Summing over them will give a value that approaches the number of observations:

```
eta.at.integration.points <- burk.res$summary.fix[1,1] +
   burk.res$summary.ran$s$mean
c(n = n, 'E(n)' = sum(st.vol * exp(eta.at.integration.points)))
##       n   E(n)
## 188.0 144.7
```

The posterior marginal distributions for the intercept and the other parameters
in the model have been plotted in Figure 8.4.

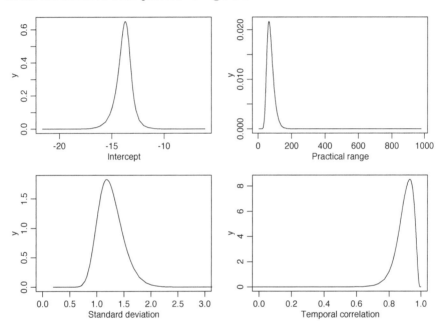

FIGURE 8.4 Intercept and random field parameters posterior marginal
distributions.

The projection over a grid for each time knot can be computed as:

```
r0 <- diff(range(burbdy[, 1])) / diff(range(burbdy[, 2]))
prj <- inla.mesh.projector(mesh.s, xlim = range(burbdy[, 1]),
   ylim = range(burbdy[, 2]), dims = c(100, 100 / r0))
ov <- over(SpatialPoints(prj$lattice$loc), domainSP)
m.prj <- lapply(1:k, function(j) {
   r <- inla.mesh.project(prj,
     burk.res$summary.ran$s$mean[1:m + (j - 1) * m])
   r[is.na(ov)] <- NA
```

```
    return(r)
})
```

The fitted latent field at each time knot has been displayed in Figure 8.5. A similar plot could be produced for the standard deviation.

8.4 Large point process dataset

In this section an approach to fit a spatio-temporal log-Gaussian Cox point process model for a large dataset is shown using a simulated dataset.

8.4.1 Simulated dataset

The dataset will be simulated by drawing samples from a separable space-time intensity function. We assume that the logarithm of the intensity function is a Gaussian process. This space-time point process can be sampled in two steps. First, a sample from a separable spacetime Gaussian process is drawn. Second, the point process is sampled conditional to this realization.

The separable space-time covariance assumed considers a Matérn covariance for space and the Exponential for time. In this case the temporal correlation at lag δt is $\mathrm{e}^{-\theta \delta t}$. Considering this continuous process sampled at equally spaced intervals $t_1, t_2, ...,$ with $t_2 - t_1 = \delta t$, then the correlation can be expressed as $\rho = \mathrm{e}^{-\theta \delta t}$. If $\delta t = 1$ we have $\rho = \mathrm{e}^{-\theta}$. This establishes a link with the first order autoregression that we will consider in the fitting process, where ρ is the lag one correlation parameter.

The sample is drawn using the lgcp package (Taylor et al., 2013). We have to specify the parameters for the Gaussian process. These are the marginal standard deviation σ, the spatial correlation parameter ϕ (which gives us $\sqrt{\phi}$ as the spatial range in our parametrization) and the temporal correlation parameter $\theta = -\log(\rho)$. These parameters are passed to the lgcpSim() function considering the lgcppars() function.

There are two additional parameters for the lgcpSim() function which are related to the mean of the Gaussian latent process, the intercept μ and β that is used in case of covariate. We can increase μ in order to increase the intensity function and then increase the number of points in the sample. The expected number of points in the sample depends on the mean of the intensity function which is modeled by the mean of the latent field, the variance of the latent field, the size of the spatial domain and the length of the time window

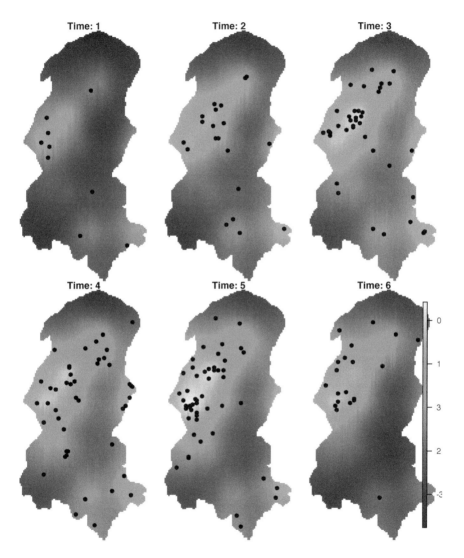

FIGURE 8.5 Fitted latent field at each time knot overlayed by the points closer in time.

as $E(N) = \exp(\mu + \sigma^2/2) * V$, where V is the area of the spatial domain times the time length.

First, the spatial domain is defined as follows:

```
x0 <- seq(0, 4 * pi, length = 15)
domain <- data.frame(x = c(x0, rev(x0), 0))
domain$y <- c(sin(x0 / 2) - 2, sin(rev(x0 / 2)) + 2, sin(0) - 2)
```

Then, it is converted into an object of the SpatialPolygons class:

```
domainSP <- SpatialPolygons(list(Polygons(list(Polygon(domain)),
   '0')))
```

The area can be computed as:

```
library(rgeos)
s.area <- gArea(domainSP)
```

We can now define the model parameters:

```
ndays <- 12

sigma <- 1
phi <- 1
range <- sqrt(8) * phi
rho <- 0.7
theta <- -log(rho)
mu <- 2

(E.N <- exp(mu + sigma^2/2) * s.area * ndays )
## [1] 7348
```

Then we use the lgcpSim() function to sample the points:

```
if(require(lgcp, quietly = TRUE)) {
    mpars <- lgcppars(sigma, phi, theta, mu - sigma^2/2)
    set.seed(1)
    xyt <- lgcpSim(
      owin = spatstat:::owin(poly = domain), tlim = c(0, ndays),
      model.parameters = mpars, cellwidth = 0.1,
      spatial.covmodel = 'matern', covpars = c(nu = 1))
    #save("xyt", file="data/xyt.RData")
```

```
} else {
  load("data/xyt.RData")
}
```

```
n <- xyt$n
```

In the previous code we have used the `require()` function to check whether the `lgcp` package can be loaded. The `lgcp` package depends on the `rpanel` package (Bowman et al., 2010) which in turn depends on the TCL/TK widget library `BWidget`. This is a system dependence, which cannot be installed from R, and may not be available on all systems by default. In case the `BWidget` library is not installed locally, the `lgcp` package will fail to install and the code above cannot be run, but the simulated data can be downloaded from the book website in order to run the examples below.

In order to fit the model, a discretization over space and over time needs to be defined. For the temporal domain, a temporal mesh based on a number of time knots will be used:

```
w0 <- 2
tmesh <- inla.mesh.1d(seq(0, ndays, by = w0))
tmesh$loc
## [1]  0  2  4  6  8 10 12
(k <- length(tmesh$loc))
## [1] 7
```

In order to consider fast computations, we lower the mesh resolution. However, it has to be tuned with the range of the spatial process. One should think about the problem of having a too coarse mesh that may not represent the Matérn field. One way to consider this is to try with a coarse mesh and look at the estimated range and then improve from there if necessary. It is better to avoid having the spatial range smaller than the mesh edge length.

The spatial mesh is defined using the domain polygon:

```
smesh <- inla.mesh.2d(boundary = inla.sp2segment(domainSP),
  max.edge = 0.75, cutoff = 0.3)
```

Figure 8.6 shows a plot of a sample of the data over time and the time knots, as well as a plot of the data over space and the spatial mesh.

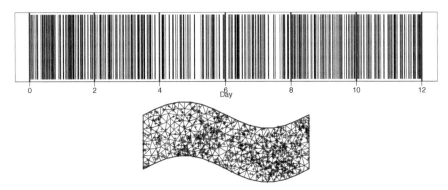

FIGURE 8.6 Time for a sample of the events (black), time knots (blue) in the upper plot. Spatial locations of another sample on the spatial domain (bottom plot).

8.4.2 Space-time aggregation

For large datasets it can be computationally demanding to fit the model. The problem is that the dimension of the model will be $n + m * k$, where n is the number of data points, m is the number of nodes in the mesh and k is the number of time knots. In this section the approach chosen to deal with the model is to aggregate the data in a way that the problem is reduced to one of dimension $2 * m * k$. So, this approach really makes sense when $n \gg m * k$.

Data will be aggregated according to the integration points to make the fitting process easier. Dual mesh polygons will also be considered, as shown in Chapter 4.

So, the first step is to find the Voronoi polygons for the mesh nodes:

```
library(deldir)
dd <- deldir(smesh$loc[, 1], smesh$loc[, 2])
tiles <- tile.list(dd)
```

Then, these are converted into a `SpatialPolygons` object, as follows:

```
polys <- SpatialPolygons(lapply(1:length(tiles), function(i) {
  p <- cbind(tiles[[i]]$x, tiles[[i]]$y)
  n <- nrow(p)
  Polygons(list(Polygon(p[c(1:n, 1), ])), i)
}))
```

The next step is to find to which polygon each data point belongs:

```
area <- factor(over(SpatialPoints(cbind(xyt$x, xyt$y)), polys),
  levels = 1:length(polys))
```

Similarly, it is necessary to find to which part of the time mesh each data point belongs:

```
t.breaks <- sort(c(tmesh$loc[c(1, k)],
  tmesh$loc[2:k - 1] / 2 + tmesh$loc[2:k] / 2))
time <- factor(findInterval(xyt$t, t.breaks),
  levels = 1:(length(t.breaks) - 1))
```

The distribution of data points on the time knots is summarized here:

```
table(time)
## time
##    1    2    3    4    5    6    7
##  657 1334  845  782  832 1022  610
```

Then, both identification index sets are used to aggregate the data:

```
agg.dat <- as.data.frame(table(area, time))
for(j in 1:2) # set time and area as integer
    agg.dat[[j]] <- as.integer(as.character(agg.dat[[j]]))
```

The resulting `data.frame` contains the area, time span and frequency of the aggregated data:

```
str(agg.dat)
## 'data.frame':     1743 obs. of  3 variables:
##  $ area: int  1 2 3 4 5 6 7 8 9 10 ...
##  $ time: int  1 1 1 1 1 1 1 1 1 1 ...
##  $ Freq: int  2 0 0 1 1 0 0 2 4 0 ...
```

The expected number of cases needs to be defined (at least) proportional to the area of the polygons times the width length of the time knots. Computing the intersection area of each polygon with the domain (show the sum) is done as follows:

```
w.areas <- sapply(1:length(tiles), function(i) {
  p <- cbind(tiles[[i]]$x, tiles[[i]]$y)
  n <- nrow(p)
```

```
pl <- SpatialPolygons(
  list(Polygons(list(Polygon(p[c(1:n, 1),])), i)))
if (gIntersects(pl, domainSP))
  return(gArea(gIntersection(pl, domainSP)))
else return(0)
})
```

A summary of the areas of the polygons is:

```
summary(w.areas)
##     Min. 1st Qu.  Median    Mean 3rd Qu.    Max.
##    0.039   0.141   0.210   0.202   0.252   0.339
```

The total sum of the weights is 50.2655 and the area of the spatial domain is:

```
s.area
## [1] 50.27
```

The time length (domain) is 12 and the width of each knot is

```
w.t <- diag(inla.mesh.fem(tmesh)$c0)
w.t
## [1] 1 2 2 2 2 2 1
```

Here, the knots at the boundaries of the time period have a lower width than the internal ones.

Since the intensity function is the number of cases per volume unit, with n cases the intensity varies about the average number of cases (intensity) by unit volume. This quantity is related to the intercept in the model. Actually, the log of it is an estimative of the intercept in the model without the space-time effect. See below:

```
i0 <- n / (gArea(domainSP) * diff(range(tmesh$loc)))
c(i0, log(i0))
## [1] 10.083  2.311
```

The space-time volume (area unit per time unit) at each polygon and time knot is:

```
e0 <- w.areas[agg.dat$area] * (w.t[agg.dat$time])
summary(e0)
```

```
##     Min. 1st Qu.  Median     Mean 3rd Qu.     Max.
##    0.039   0.222   0.336    0.346   0.466    0.679
```

8.4.3 Model fitting

The projector matrix, SPDE model object and the space-time index set definition are computed as follows:

```
A.st <- inla.spde.make.A(smesh, smesh$loc[agg.dat$area, ],
  group = agg.dat$time, mesh.group = tmesh)
spde <- inla.spde2.pcmatern(
  smesh, prior.sigma = c(1,0.01), prior.range = c(0.05,0.01))
idx <- inla.spde.make.index('s', spde$n.spde, n.group = k)
```

The data stack is defined as:

```
stk <- inla.stack(
  data = list(y = agg.dat$Freq, exposure = e0),
  A = list(A.st, 1),
  effects = list(idx, list(b0 = rep(1, nrow(agg.dat)))))
```

The formula to fit the model considers the intercept, spatial effect and temporal effect:

```
# PC prior on correlation
pcrho <- list(theta = list(prior = 'pccor1', param = c(0.7, 0.7)))
# Model formula
formula <- y ~ 0 + b0 +
  f(s, model = spde, group = s.group,
    control.group = list(model = 'ar1', hyper = pcrho))
```

Finally, model fitting is carried out:

```
res <- inla(formula, family = 'poisson',
  data = inla.stack.data(stk), E = exposure,
  control.predictor = list(A = inla.stack.A(stk)),
  control.inla = list(strategy ='adaptive'))
```

The value of μ and the intercept summary can be obtained as follows:

```
cbind(True = mu, res$summary.fixed[, 1:6])
##      True  mean      sd 0.025quant 0.5quant 0.975quant  mode
## b0    2 2.045 0.177       1.696     2.045      2.396 2.044
```

The expected number of cases at each integration point can be used to compute the total expected number of cases (Est. N below), as:

```
eta.i <- res$summary.fix[1, 1] + res$summary.ran$s$mean
c('E(N)' = E.N, 'Obs. N' = xyt$n,
  'Est. N' = sum(rep(w.areas, k) *
     rep(w.t, each = smesh$n) * exp(eta.i)))
##    E(N) Obs. N Est. N
##   7348   6082   5726
```

A summary for the hyperparameters can be obtained with this R code:

```
cbind(True = c(range, sigma, rho),
  res$summary.hyperpar[, c(1, 2, 3, 5)])
##                     True   mean      sd 0.025quant 0.975quant
## Range for s        2.828 2.4240 0.19804     2.0651     2.8427
## Stdev for s        1.000 0.6812 0.03794     0.6105     0.7594
## GroupRho for s     0.700 0.5332 0.04507     0.4421     0.6192
```

The spatial surface at each time knot can be computed as well:

```
r0 <- diff(range(domain[, 1])) / diff(range(domain[, 2]))
prj <- inla.mesh.projector(smesh, xlim = bbox(domainSP)[1, ],
  ylim = bbox(domainSP)[2, ], dims = c(r0 * 200, 200))
g.no.in <- is.na(over(SpatialPoints(prj$lattice$loc), domainSP))
t.mean <- lapply(1:k, function(j) {
  z.j <- res$summary.ran$s$mean[idx$s.group == j]
  z <- inla.mesh.project(prj, z.j)
  z[g.no.in] <- NA
  return(z)
})
```

Figure 8.7 shows the predicted surfaces at each time knot.

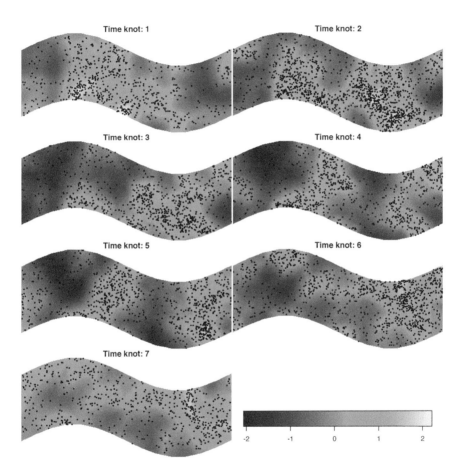

FIGURE 8.7 Spatial surface fitted at each time knot overlayed by the point pattern formed by the points nearest to each time knot.

8.5   Accumulated rainfall: Hurdle Gamma model

For some applications it is possible to have the outcome be zero or a positive number. Common examples are fish biomass and accumulated rainfall. In this case one can build a model that accommodates the zero and positive outcome considering a combination of two likelihoods, one to model the occurrence and another to model the amount. One case is when considering the Bernoulli distribution for the occurrence and the Gamma for the amount. The advantage of having this two-part model is that we can model the probability of rain and the rainfall amount separately. It may be the case that some terms in each part can be shared.

8.5.1   The model

We will consider the daily rainfall data considered in Section 2.8. Let

$$
z_{i,t} = \begin{cases} 1, & \text{if it has rained at location } \mathbf{s}_i \text{ and time } t \\[2ex] 0, & \text{otherwise} \end{cases} \tag{8.1}
$$

and the rainfall amount as

$$
y_{i,t} = \begin{cases} \text{NA}, & \text{if it did not rain at} \\ & \text{location } \mathbf{s}_i \text{ and time } t \\ \text{rainfall amount at location } \mathbf{s}_i \text{ and time } t, & \text{otherwise} \end{cases} \tag{8.2}
$$

We then define a likelihood for each outcome. We choose to set a Bernoulli distribution for z_i and a Gamma distribution for y_i:

$$
z_{i,t} \sim \text{Binomial}(\pi_{i,t}, n_{i,t} = 1) \quad \text{and} \quad y_{i,t} \sim \text{Gamma}(a_{i,t}, b_{i,t}). \tag{8.3}
$$

This setting is equivalent to consider a Hurdle-Gamma model where we have the expected value of the rainfall as $p_{i,t} + (1 - p_{i,t})\mu_{i,t}$ where $\mu_{i,t}$ is the expected value for the Gamma part. Next, we define the model for $p_{i,t}$ and $\mu_{i,t}$.

For the occurrence, the model is for the linear predictor as the *logit* of the probability as usual for the Bernoulli, specified as

$$
\text{logit}(\pi_{i,t}) = \alpha^z + \xi_{i,t} \tag{8.4}
$$

with α^z being the intercept and ξ_i coming from a space-time random effect, i.e. a GF modeled through the SPDE approach.

The parameterization of the Gamma distribution in R-INLA considers that $E(y) = \mu = a/b$ and $Var(y) = a/b^2 = 1/\tau$, where τ is the precision parameter. The linear predictor is defined on $\log(\mu)$ and we have

$$\log(\mu_{i,t}) = \alpha^y + \beta\xi_{i,t} + u_{i,t} \tag{8.5}$$

where α^y is the intercept and β the scaling parameter for $\xi_{i,t}$, which is the space-time effect considered for the occurrence probability, which is being shared in the model for the rainfall amount. The linear predictor affects both the $E(y)$ and $Var(y)$ because $a = b\mu$ and then $a/b^2 = \mu/b$.

Notice that $\xi_{i,t}$ will be computed as $\xi_{i,t}\mathbf{A}\xi_0$, where ξ_0 is the space-time process at the mesh nodes and time points and $\mathbf{A}$ is the correspondent space-time projector matrix. This is similar for $u_{i,t}$.

We will consider the model in Cameletti et al. (2013) for both ξ and $\mathbf{u}$. However, we will consider the PC-prior for each one of the three parameters. Thus for the marginal standard deviation and the spatial range we consider the prior proposed in Fuglstad et al. (2018). We set it such that the standard deviation median is 0.5, $P(\sigma > 0.5 = 0.5)$.

```
psigma <- c(0.5, 0.5)
```

For the practical range we consider the size of the Parana state. First we load the data PRprec which also load the Parana border, PRborder.

```
data(PRprec)
head(PRborder,2)
##      Longitude Latitude
## [1,]   -54.61   -25.45
## [2,]   -54.60   -25.43
```

We consider that the coordinates are in longitude and latitude and project them into UTM with units in kilometers as follows:

```
border.ll <- SpatialPolygons(list(Polygons(list(
  Polygon(PRborder)), '0')),
  proj4string = CRS("+proj=longlat +datum=WGS84"))
border <- spTransform(border.ll,
  CRS("+proj=utm +units=km +zone=22 +south"))
bbox(border)
##      min    max
## x    136   799.9
## y   7045  7509.6
```

```
apply(bbox(border), 1, diff)
##      x     y
## 663.9 464.7
```

We have that the Paraná state is around 663.8711 kilometers width by 464.7481 kilometers height. The PC-prior for the practical range is built considering the probability of the practical range being less than a chosen distance. We chose to set the prior considering the median as 100 kilometers.

```
prange <- c(100, 0.5)
```

For the temporal correlation parameter, the first order autoregression parameter, we also consider the PC-prior framework, as in Simpson et al. (2016). We choose to have the correlation one as the base model and set the prior considering $P(\rho > 0.5)=0.7$ as follows:

```
rhoprior <- list(theta = list(prior = 'pccor1',
  param = c(0.5, 0.7)))
```

For the shared parameter β we can set a prior based on some knowledge about the correlation between the rain occurrence and the rainfall amount. We assume a N(0, 1) prior for this parameter as follows

```
bprior <- list(prior = 'gaussian', param = c(0,1))
```

We also have a likelihood parameter to set prior on and, again, we consider the PC-prior framework. Thus we set the prior on the precision choosing a value for λ. We consider it equals one as follows:

```
pcgprior <- list(prior = 'pc.gamma', param = 1)
```

8.5.2 Rainfall data in Paraná

In this section we consider the rainfall data introduced in Section 2.8. In this data we have the longitude in the first column, the latitude in the second, the altitude in the third and from the fourth column we have the data for each day, as we can see below:

```
PRprec[1:3, 1:8]
##    Longitude Latitude Altitude d0101 d0102 d0103 d0104 d0105
```

```
## 1     -50.87    -22.85       365     0     0      0    0.0      0
## 3     -50.77    -22.96       344     0     1      0    0.0      0
## 4     -50.65    -22.95       904     0     0      0    3.3      0
loc.ll <- SpatialPoints(PRprec[,1:2], border.ll@proj4string)
loc <- spTransform(loc.ll, border@proj4string)
```

We will consider the first 8 days of data for illustration. The two response variables z_i and y_i are defined as follows. First we define the occurrence variable

```
m <- 8
days.idx <- 3 + 1:m
z <- as.numeric(PRprec[, days.idx] > 0)
table(z)
## z
##    0    1
## 3153 1719
```

The rainfall is then defined as

```
y <- ifelse(z == 1, unlist(PRprec[, days.idx]), NA)
table(is.na(y))
##
## FALSE   TRUE
##  1719   3209
```

8.5.3   Fitting the model

We have to build a mesh in order to define the SPDE model. We consider all the gauges' locations in the following code:

```
mesh <- inla.mesh.2d(loc, max.edge = 200, cutoff = 35,
  offset = 150)
```

And we have the resulting mesh in Figure 8.8.

The SPDE model is defined through

```
spde <- inla.spde2.pcmatern(
  mesh, prior.range = prange, prior.sigma = psigma)
```

and the corresponding spacetime predictor matrix is given by

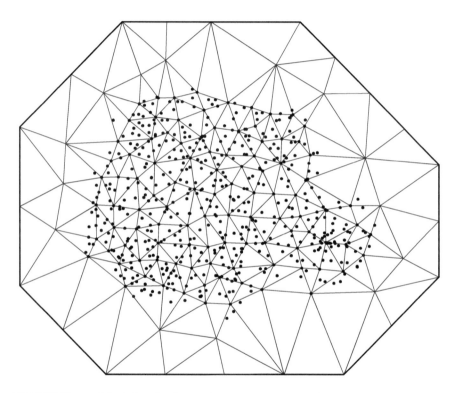

FIGURE 8.8 Mesh for the Paraná state with 138 nodes. Black points denote the 616 rain gauges.

```
n <- nrow(PRprec)
stcoords <- kronecker(matrix(1, m, 1), coordinates(loc))
A <- inla.spde.make.A(mesh = mesh, loc = stcoords,
  group = rep(1:m, each = n))
dim(A) == (m * c(n, spde$n.spde)) # Check that dimensions match
## [1] TRUE TRUE
```

We need to define the space-time indices ξ in both the linear predictors.

```
field.z.idx <- inla.spde.make.index(name = 'x',
  n.spde = spde$n.spde, n.group = m)
field.zc.idx <- inla.spde.make.index(name = 'xc',
  n.spde = spde$n.spde, n.group = m)
field.y.idx <- inla.spde.make.index(name = 'u',
  n.spde = spde$n.spde, n.group = m)
```

The next step is to organize the data into stacks. First, we create a data stack for the occurrence data bearing in mind that we have the amount data. So, we have the occurrence in the first column of a two-column matrix and the amount of rainfall in the second column:

```
stk.z <- inla.stack(
  data = list(Y = cbind(as.vector(z), NA), link = 1),
  A = list(A, 1),
  effects = list(field.z.idx, z.intercept = rep(1, n * m)),
  tag = 'zobs')

stk.y <- inla.stack(
  data = list(Y = cbind(NA, as.vector(y)), link = 2),
  A = list(A, 1),
  effects = list(c(field.zc.idx, field.y.idx),
  y.intercept = rep(1, n * m)),
  tag = 'yobs')
```

It is useful to have a stack for prediction at the mesh nodes so it will be easy to map the predicted values later:

```
stk.zp <- inla.stack(
  data = list(Y = matrix(NA, ncol(A), 2), link = 1),
  effects = list(field.z.idx, z.intercept = rep(1, ncol(A))),
  A = list(1, 1),
  tag = 'zpred')
```

```
stk.yp <- inla.stack(
  data = list(Y = matrix(NA, ncol(A), 2), link = 2),
  A = list(1, 1),
  effects = list(c(field.zc.idx, field.y.idx),
    y.intercept = rep(1, ncol(A))),
  tag = 'ypred')
```

We join all the data stacks:

```
stk.all <- inla.stack(stk.z, stk.y, stk.zp, stk.yp)
```

We now set some parameters to supply for the `inla()` function. The prior for the precision parameter of the Gamma likelihood will go in a list for control family arguments.

```
cff <- list(list(), list(hyper = list(theta = pcgprior)))
```

Note that the empty list above, i.e., `list()`, is required and it could be used to pass additional arguments to the Binomial likelihood in the model.

For having a fast approximation of the marginals we use the adaptive approximation strategy (by setting `strategy = 'adaptive'` below). This strategy mostly uses the Gaussian approximation to avoid the second Laplace approximation in the INLA algorithm, but applies the default strategy for fixed effects and random effects with a length $\leq$ `adaptive.max` (see `?control.inla`). Additionally we choose to not integrate over the hyperparameters by choosing the Empirical Bayes estimation as `int.strategy = 'eb'`. These options will be passed in argument `control.inla` to function `inla()` when fitting the model and are defined now:

```
cinla <- list(strategy = 'adaptive', int.strategy = 'eb')
```

We also consider not to return the marginal distribution for the latent field. Thus we set

```
cres <- list(return.marginals.predictor = FALSE,
  return.marginals.random = FALSE)
```

We can define the model formula for the model specified. We use the `spde` object to define the model of the space-time component together with the definition of the prior for the AR(1) temporal dynamics. In order to define the β parameter of the shared space-time component we set `fixed = FALSE` to estimate β and insert its prior. In order to achieve fewer iterations during the optimization over the posterior for the hyperparameters we set initial values near the optimal ones as we have run this model previously, and restart the optimization from there. The joint model with the shared space-time component is then fitted as follows:

```
cg <- list(model = 'ar1', hyper = rhoprior)
formula.joint <- Y ~ -1 + z.intercept + y.intercept +
  f(x, model = spde, group = x.group, control.group = cg) +
  f(xc, copy = "x", fixed = FALSE, group = xc.group,
    hyper = list(theta = bprior)) +
  f(u, model = spde, group = u.group, control.group = cg)

# Initial values of parameters
ini.jo <- c(-0.047, 5.34, 0.492, 1.607, 4.6, -0.534, 1.6, 0.198)

res.jo <- inla(formula.joint, family = c("binomial", "gamma"),
  data = inla.stack.data(stk.all), control.family = cff,
  control.predictor = list(A = inla.stack.A(stk.all),
    link = link),
  control.compute = list(dic = TRUE, waic = TRUE, cpo = TRUE,
    config = TRUE),
  control.results = cres, control.inla = cinla,
  control.mode = list(theta = ini.jo, restart = TRUE))
```

The model without the shared space-time component is fitted as follows:

```
formula.zy <- Y ~ -1 + z.intercept + y.intercept +
  f(x, model = spde, group = x.group, control.group = cg) +
  f(u, model = spde, group = u.group, control.group = cg)

# Initial values of parameters
ini.zy <- c(-0.05, 5.3, 0.5, 1.62, 4.65, -0.51, 1.3)

res.zy <- inla(formula.zy, family = c("binomial", "gamma"),
  data = inla.stack.data(stk.all), control.family = cff,
  control.predictor = list(A =inla.stack.A(stk.all),
    link = link),
  control.compute=list(dic = TRUE, waic = TRUE, cpo = TRUE,
    config = TRUE),
  control.results = cres, control.inla = cinla,
  control.mode = list(theta = ini.zy, restart = TRUE))
```

Alternatively, the model with only the shared component is fitted as follows:

```
formula.sh <- Y ~ -1 + z.intercept + y.intercept +
  f(x, model = spde, group = x.group, control.group = cg) +
  f(xc, copy = "x", fixed = FALSE, group = xc.group)

# Initial values of parameters
ini.sh <- c(-0.187, 5.27, 0.47, 1.47, 0.17)

res.sh <- inla(formula.sh, family = c("binomial", "gamma"),
  data = inla.stack.data(stk.all), control.family = cff,
  control.predictor = list(
    A = inla.stack.A(stk.all), link = link),
  control.compute = list(dic = TRUE, waic = TRUE, cpo = TRUE,
    config = TRUE),
  control.results = cres, control.inla = cinla,
  control.mode = list(theta = ini.sh, restart = TRUE))
```

Sometimes, the CPO is not computed automatically for all the observations.
In this case we can use the `inla.cpo()` function to manually compute it.

```
sum(res.jo$cpo$failure, na.rm = TRUE)
sum(res.zy$cpo$failure, na.rm = TRUE)
sum(res.sh$cpo$failure, na.rm = TRUE)

res.jo <- inla.cpo(res.jo, verbose = FALSE)

res.zy <- inla.cpo(res.zy, verbose = FALSE)
res.sh <- inla.cpo(res.sh, verbose = FALSE)
```

We can now perform a model comparison. This can be done using the marginal
likelihood, DIC, WAIC or CPO. Because we have two outcomes, we need to
account for this with care. The DIC, WAIC and CPO are computed for each
observation. Thus, we can sum it for each outcome as follows:

```
getfit <- function(r) {
  fam <- r$dic$family
  data.frame(dic = tapply(r$dic$local.dic, fam, sum),
    waic = tapply(r$waic$local.waic, fam, sum),
    cpo = tapply(r$cpo$cpo, fam,
      function(x) - sum(log(x), na.rm = TRUE)))
}
rbind(separate = getfit(res.jo),
  joint = getfit(res.zy),
  oshare = getfit(res.sh))[c(1, 3, 5, 2, 4, 6),]
```

```
##                     dic   waic   cpo
## separate.1   5094   5082  2542
## joint.1      5097   5084  2543
## oshare.1     5101   5088  2545
## separate.2  11281  11297  5693
## joint.2     11294  11310  5705
## oshare.2    11457  11458  5729
```

and we can see that the separate model fits slightly better.

8.5.4 Visualizing some results

We extract the useful indices for later use, one for each outcome at the observation locations and one for each outcome at the mesh locations, for all time points:

```
idx.z <- inla.stack.index(stk.all, 'zobs')$data
idx.y <- inla.stack.index(stk.all, 'yobs')$data
idx.zp <- inla.stack.index(stk.all, 'zpred')$data
idx.yp <- inla.stack.index(stk.all, 'ypred')$data
```

It may be useful to show maps of the space-time effect at each time, or the probability of rain or the expected value of rainfall. In order to compute it we do need to have the projector from the mesh nodes to a fine grid:

```
wh <- apply(bbox(border), 1, diff)
nxy <- round(300 * wh / wh[1])
pgrid <- inla.mesh.projector(mesh, xlim = bbox(border)[1, ],
  ylim = bbox(border)[2, ], dims = nxy)
```

It is better to discard the values interpolated outside the border. Thus we identify those pixels which are outside of the Paraná border:

```
ov <- over(SpatialPoints(pgrid$lattice$loc,
  border@proj4string), border)
id.out <- which(is.na(ov))
```

Figure 8.9 shows the posterior mean of the probability of rain at each time known. It has been produced with the following code:

```
stpred <- matrix(res.jo$summary.fitted.values$mean[idx.zp],
  spde$n.spde)
```

```
par(mfrow = c(4, 2), mar =c(0, 0, 0, 0))
for (j in 1:m) {
  pj <- inla.mesh.project(pgrid, field = stpred[, j])
  pj[id.out] <- NA
  book.plot.field(list(x = pgrid$x, y = pgrid$y, z = pj),
    zlim = c(0, 1))
}
```

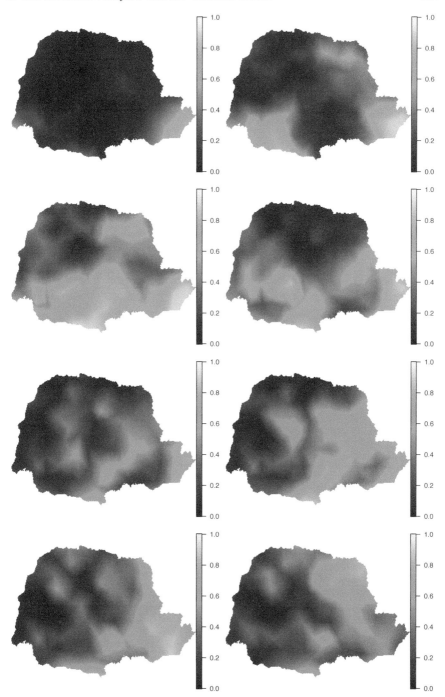

FIGURE 8.9 Posterior mean of the probability of rain at each time knot. Time flows from top to bottom and left to right.

A

List of symbols and notation

$\mathbf{s} = (s_1, s_2)$ location in space.

$\mathbf{s}_i = (s_{1i}, s_{2i})$ location in space of observation i.

D is the domain of the study region.

$\Re^d$ is a real coordinate space of dimension d.

$U(\mathbf{s})$ is a stochastic process.

$u(\mathbf{s}_i)$ is a realization of $U(\mathbf{s})$ at location $\mathbf{s}_i$.

n is the number of sampling locations.

h is a distance (usually, between two points).

y_i is an observation at location $\mathbf{s}_i$.

μ is the mean or intercept of the model.

μ_i is the mean of the spatial process at location $\mathbf{s}_i$.

e_i is the error term.

σ_e^2 is the variance of the error term.

Σ is a variance-covariance matrix.

$\mathbf{F}_i$ is a matrix of covariates at location $\mathbf{s}_i$.

β is a vector of coefficients of covariates.

β_j is the coefficient of covariate j.

κ is the scale parameter of Matérn covariance.

ν is the smoothness parameter of Matérn covariance.

$\| \, . \, \|$ denotes the Euclidean distance.

$K_\nu(\cdot)$ is the modified Bessel function of the second kind.

σ_u^2 is the marginal variance of Matérn process.

$\mathbf{u}$ is a sample from a Matérn process.

$\mathbf{z}$ is a vector of n samples from a standard Gaussian distribution.

R is the Cholesky decomposition of the covariance.

$E(\cdot)$ denotes the expectation.

$Cor(\cdot,\cdot)$ denotes the correlation.

$Cor_M(\cdot,\cdot)$ denotes the correlation of a Matérn process.

I is the identity matrix.

$\lambda(s)$ is the intensity of a point process at location s.

$S(s)$ is a continuous spatial Gaussian process (usually with a Matérn covariance).

B

Packages used in the book

We have included below the information about the packages used when compiling this book. This will be useful when trying to reproduce the results and figures. Different versions of R and the packages shown below may produce slightly different results. Furthermore, the architecture may also cause minor differences in the results of model fitting.

```
## R version 3.5.1 (2018-07-02)
## Platform: x86_64-apple-darwin15.6.0 (64-bit)
## Running under: macOS High Sierra 10.13.6
##
## Matrix products: default
## BLAS: /R/Versions/3.5/Resources/lib/libRblas.0.dylib
## LAPACK: /R/Versions/3.5/Resources/lib/libRlapack.dylib
##
## attached base packages:
## [1] grid      parallel  stats     graphics  grDevices utils
## [7] datasets  methods   base
##
## other attached packages:
##  [1] survival_2.42-6     scales_1.0.0
##  [3] splancs_2.01-40     spelling_1.2
##  [5] spdep_0.8-1         spData_0.2.9.4
##  [7] spatstat_1.56-1     rpart_4.1-13
##  [9] nlme_3.1-137        spatstat.data_1.3-1
## [11] rgeos_0.3-28        rgdal_1.3-4
## [13] osmar_1.1-7         geosphere_1.5-7
## [15] RCurl_1.95-4.11     bitops_1.0-6
## [17] XML_3.98-1.16       maptools_0.9-3
## [19] mapdata_2.3.0       latticeExtra_0.6-28
## [21] lattice_0.20-35     inlabru_2.1.9
## [23] ggplot2_3.0.0       gridExtra_2.3
## [25] evd_2.3-3           deldir_0.1-15
## [27] RColorBrewer_1.1-2  viridisLite_0.3.0
## [29] fields_9.6          maps_3.3.0
## [31] spam_2.2-0          dotCall64_1.0-0
## [33] knitr_1.20          INLA_18.09.24
```

```
## [35] sp_1.3-1              Matrix_1.2-14
##
## loaded via a namespace (and not attached):
##  [1] splines_3.5.1         gtools_3.8.1
##  [3] assertthat_0.2.0      expm_0.999-3
##  [5] LearnBayes_2.15.1     yaml_2.2.0
##  [7] pillar_1.3.0          backports_1.1.2
##  [9] glue_1.3.0            digest_0.6.17
## [11] polyclip_1.9-1        colorspace_1.3-2
## [13] htmltools_0.3.6       plyr_1.8.4
## [15] pkgconfig_2.0.2       gmodels_2.18.1
## [17] bookdown_0.7          purrr_0.2.5
## [19] gdata_2.18.0          tensor_1.5
## [21] spatstat.utils_1.9-0  tibble_1.4.2
## [23] mgcv_1.8-24           withr_2.1.2
## [25] lazyeval_0.2.1        magrittr_1.5
## [27] crayon_1.3.4          evaluate_0.11
## [29] MASS_7.3-50           foreign_0.8-71
## [31] tools_3.5.1           stringr_1.3.1
## [33] munsell_0.5.0         bindrcpp_0.2.2
## [35] compiler_3.5.1        rlang_0.2.2
## [37] goftest_1.1-1         rmarkdown_1.10
## [39] boot_1.3-20           gtable_0.2.0
## [41] abind_1.4-5           R6_2.2.2
## [43] dplyr_0.7.6           bindr_0.1.1
## [45] rprojroot_1.3-2       stringi_1.2.4
## [47] Rcpp_0.12.18          coda_0.19-2
## [49] tidyselect_0.2.4      xfun_0.3
```

End.

Bibliography

Abrahamsen, P. (1997). A review of Gaussian random fields and correlation functions. Norwegian Computing Center report No. 917.

Andersen, P. K. and R. D. Gill (1982). Cox's regression model for counting processes: A large sample study. *The Annals of Statistics 10*(4), 1100–1120.

Assunção, J. J., D. Gamerman, and R. M. Assunção (1999). Regional differences in factor productivities of Brazilian agriculture: A space-varying parameter approach. Technical report, Universidade Federal do Rio de Janeiro, Statistical Laboratory.

Assunção, R. M., J. E. Potter, and S. M. Cavenaghi (2002). A Bayesian space varying parameter model applied to estimating fertility schedules. *Statistics in Medicine 21*, 2057–2075.

Baddeley, A., E. Rubak, and R. Turner (2015). *Spatial Point Patterns: Methodology and Applications with R*. Boca Raton, FL: Chapman & Hall/CRC.

Bailey, T. C. and A. C. Gatrell (1995). *Interactive Spatial Data Analysis*. Harlow, UK: Longman Scientific & Technical.

Bakka, H. (2018). How to solve the stochastic partial differential equation that gives a Matérn random field using the finite element method. *arXiv preprint arXiv:1803.03765*.

Bakka, H., H. Rue, G.-A. Fuglstad, A. Riebler, D. Bolin, E. Krainski, D. Simpson, and F. Lindgren (2018). Spatial modelling with R-INLA: A review. *WIREs Comput Stat 10*(6), 1–24.

Bakka, H., J. Vanhatalo, J. Illian, D. Simpson, and H. Rue (2016, August). Non-stationary Gaussian models with physical barriers. *ArXiv e-prints*.

Banerjee, S., B. P. Carlin, and A. E. Gelfand (2014). *Hierarchical Modeling and Analysis for Spatial Data* (2nd ed.). Boca Raton, FL: Chapman & Hall/CRC.

Becker, R. A., A. R. Wilks, and R. Brownrigg (2016). *mapdata: Extra Map Databases*. R package version 2.2-6.

Besag, J. (1981). On a system of two-dimensional recurrence equations. *J. R. Statist. Soc. B 43*(3), 302–309.

Bivand, R., T. Keitt, and B. Rowlingson (2017). *rgdal: Bindings for the 'Geospatial' Data Abstraction Library.* R package version 1.2-16.

Bivand, R., J. Nowosad, and R. Lovelace (2018). *spData: Datasets for Spatial Analysis.* R package version 0.2.7.0.

Bivand, R. and G. Piras (2015). Comparing implementations of estimation methods for spatial econometrics. *Journal of Statistical Software 63*(18), 1–36.

Bivand, R. and C. Rundel (2017). *rgeos: Interface to Geometry Engine - Open Source ('GEOS').* R package version 0.3-26.

Bivand, R. S., E. J. Pebesma, and V. Gomez-Rubio (2013). *Applied Spatial Data Analysis with R* (2 ed.). Springer, NY.

Blangiardo, M. and M. Cameletti (2015). *Spatial and SpatioTemporal Bayesian Models with R-INLA.* Chichester, UK: John Wiley & Sons, Ltd.

Bowman, A. W., I. Gibson, E. M. Scott, and E. Crawford (2010). Interactive teaching tools for spatial sampling. *Journal of Statistical Software 36*(13), 1–17.

Box, G. E. and N. R. Draper (2007). *Response Surfaces, Mixtures, and Ridge Analyses* (2nd ed.). Hoboken, NJ: Wiley-Interscience.

Brenner, S. C. and R. Scott (2007). *The Mathematical Theory of Finite Element Methods* (3rd ed.). Springer, New York.

Cameletti, M., F. Lindgren, D. Simpson, and H. Rue (2013). Spatio-temporal modeling of particulate matter concentration through the spde approach. *Advances in Statistical Analysis 97*(2), 109–131.

Carlin, B. P. and T. A. Louis (2008). *Bayesian Methods for Data Analysis* (3rd ed.). Boca Raton, FL: Chapman & Hall/CRC.

Casson, E. and S. Coles (1999). Spatial regression models for extremes. *Extremes 1*(4), 449–468.

Castro-Camilo, D. and M. de Carvalho (2017). Spectral density regression for bivariate extremes. *Stochastic Environmental Research and Risk Assessment 31*(7), 1603–1613.

Castro-Camilo, D., M. de Carvalho, and J. Wadsworth (2018). Time-varying extreme value dependence with application to leading European stock markets. *Annals of Applied Statistics 12*(1), 283–309.

Chang, W., J. Cheng, J. J. Allaire, Y. Xie, and J. McPherson (2018). *shiny: Web Application Framework for R.* R package version 1.1.0.

Ciarlet, P. G. (1978). *The Finite Element Method for Elliptic Problems.* Amsterdam: North-Holland.

Coles, S. (2001). *An Introduction to Statistical Modeling of Extreme Values*, Volume 208. London: Springer-Verlag.

Cooley, D., D. Nychka, and P. Naveau (2007). Bayesian spatial modeling of extreme precipitation return levels. *Journal of the American Statistical Association 102*(479), 824–840.

Cressie, N. (1993). *Statistics for Spatial Data*. New York: Wiley. 990p.

Cressie, N. and C. K. Wikle (2011). *Statistics for Spatio-Temporal Data*. Hoboken, NJ: John Wiley & Sons, Inc.

Davison, A. C. and R. L. Smith (1990). Models for exceedances over high thresholds (with discussion). *Journal of the Royal Statistical Society: Series B*, 393–442.

de Berg, M., O. Cheong, M. van Kreveld, and M. Overmars (2008). *Computational Geometry* (3rd ed.). Berlin Heidelberg: Springer-Verlag.

Diggle, P. J. (2013). *Statistical Analysis of Spatial and Spatio-Temporal Point Patterns* (3rd ed.). Boca Raton, FL: Chapman & Hall/CRC Press.

Diggle, P. J. (2014). *Statistical Analysis of Spatial and Spatio-Temporal Point Patterns* (3rd ed.). Boca Raton, FL: CRC Press, Taylor & Francis Group.

Diggle, P. J., R. Menezes, and T.-L. Su (2010). Geostatistical inference under preferential sampling. *Journal of the Royal Statistical Society: Series C (Applied Statistics) 59*(2), 191–232.

Diggle, P. J. and P. J. Ribeiro Jr. (2007). *Model-Based Geostatistics*. Springer Series in Statistics. Hardcover. 230p.

Eugster, M. J. A. and T. Schlesinger (2013, June). osmar: OpenStreetMap and R. *The R Journal 5*(1), 53–63.

Fuglstad, G.-A., D. Simpson, F. Lindgren, and H. Rue (2018). Constructing priors that penalize the complexity of Gaussian random fields. *Journal of the American Statistical Association to appear*.

Gamerman, D., A. R. B. Moreira, and H. Rue (2003). Space-varying regression models: specifications and simulation. *Computational Statistics & Data Analysis - Special Issue: Computational Econometrics 42*(3), 513–533.

Gelfand, A. E., H. Kim, C. F. Sirmans, and S. Banerjee (2003). Spatial modeling with spatially varying coefficient processes. *Journal of the American Statistical Association 98*(462), 387–396.

Gelfand, A. E., A. M. Schmidt, and C. F. Sirmans (2002). Multivariate spatial process models: conditional and unconditional Bayesian approaches using coregionalization. *Center for Real Estate and Urban Economic Studies, University of Connecticut*.

Gilks, W. R., S. Richardson, and D. Spiegelhalter (1996). *Markov Chain Monte Carlo in Practice*. Boca Raton, FL: Chapman & Hall.

Gómez-Rubio, V. (2019). *Bayesian Inference with INLA*. Chapman & Hall/CRC.

Guibas, L. J., D. E. Knuth, and M. Sharir (1992). Randomized incremental construction of Delaunay and Voronoi diagrams. *Algorithmica 7*, 381–413.

Haining, R. (2003). *Spatial Data Analysis: Theory and Practice*. Cambridge University Press.

Henderson, R., S. Shimakura, and D. Gorst (2003). Modeling spatial variation in leukemia survival data. *JASA 97*(460), 965–972.

Holford, T. R. (1980). The analysis of rates and of survivorship using log-linear models. *Biometrics 36*, 299–305.

Ibrahim, J. G., M.-H. Chen, and D. Sinha (2001). *Bayesian Survival Analysis*. New York: Springer-Verlag.

Illian, J., A. Penttinen, H. Stoyan, and D. Stoyan (2008). *Statistical Analysis and Modelling of Spatial Point Patterns*. Chichester, UK: Wiley & Sons.

Illian, J. B., S. H. Sørbye, and H. Rue (2012). A toolbox for fitting complex spatial point process models using integrated nested Laplace approximation (INLA). *Annals of Applied Statistics 6*(4), 1499–1530.

Ingebrigtsen, R., F. Lindgren, and I. Steinsland (2014). Spatial models with explanatory variables in the dependence structure. *Spatial Statistics 8*, 20–38.

Knorr-Held, L. and H. Rue (2002). On block updating in Markov random field models for disease mapping. *Scandinavian Journal of Statistics 20*, 597–614.

Laird, N. and D. Olivier (1981). Covariance analysis of censored survival data using log-linear analysis techniques. *Journal of the American Statistical Association 76*, 231–240.

Lindgren, F. (2012). Continuous domain spatial models in R-INLA. *The ISBA Bulletin 19*(4), 14–20.

Lindgren, F. and H. Rue (2015). Bayesian spatial and spatio-temporal modelling with R-INLA. *Journal of Statistical Software 63*(19).

Lindgren, F., H. Rue, and J. Lindström (2011). An explicit link between Gaussian fields and Gaussian Markov random fields: the stochastic partial differential equation approach (with discussion). *J. R. Statist. Soc. B 73*(4), 423–498.

Little, R. J. A. and D. B. Rubin (2002). *Statistical Analysis with Missing Data* (2nd ed.). Hoboken, NJ: John Wiley and Sons.

Marshall, E. C. and D. J. Spiegelhalter (2003). Approximate cross-validatory predictive checks in disease mapping models. *Statistics in Medicine 22*(10), 1649–1660.

Martino, S., R. Akerkar, and H. Rue (2010). Approximate Bayesian inference for survival models. *Scandinavian Journal of Statistics 28*(3), 514–528.

Martins, T. G., D. Simpson, F. Lindgren, and H. Rue (2013). Bayesian computing with INLA: New features. *Computational Statistics & Data Analysis 67*(0), 68–83.

Møller, J., A. R. Syversveen, and R. P. Waagepetersen (1998). Log Gaussian Cox processes. *Scandinavian Journal of Statistics 25*, 451–482.

Møller, J. and R. P. Waagepetersen (2003). *Statistical Inference and Simulation for Spatial Point Processes*. Boca Raton, FL: Chapman & Hall/CRC.

Muff, S., A. Riebler, L. Held, H. Rue, and P. Saner (2014). Bayesian analysis of measurement error models using integrated nested Laplace approximations. *Journal of the Royal Statistical Society: Series C 64*(2), 231–252.

OpenStreetMap contributors (2018). Planet dump retrieved from https://planet.osm.org . `https://www.openstreetmap.org`.

Opitz, T., R. Huser, H. Bakka, and H. Rue (2018). INLA goes extreme: Bayesian tail regression for the estimation of high spatio-temporal quantiles. *Extremes 21*(3), 441–462.

Petris, G., S. Petroni, and P. Campagnoli (2009). *Dynamic Linear Models with R*. New York: Springer-Verlag.

Pettit, L. I. (1990). The conditional predictive ordinate for the normal distribution. *Journal of the Royal Statistical Society: Series B (Methodological) 52*(1), 175–184.

Quarteroni, A. M. and A. Valli (2008). *Numerical Approximation of Partial Differential Equations* (2nd ed.). New York: Springer.

Ribeiro Jr., P. J. and P. J. Diggle (2001, June). geoR: a package for geostatistical analysis. *R-NEWS 1*(2), 14–18. ISSN 1609-3631.

Rowlingson, B. and P. Diggle (2017). *splancs: Spatial and Space-Time Point Pattern Analysis*. R package version 2.01-40.

Rowlingson, B. S. and P. J. Diggle (1993). Splancs: Spatial point pattern analysis code in s-plus. *Computers & Geosciences 19*(5), 627 – 655.

Rozanov, J. A. (1977). Markov random fields and stochastic partial differential equations. *Math. USSR Sbornik 32*(4)), 515–534.

Rue, H. and L. Held (2005). *Gaussian Markov Random Fields: Theory and*

Applications. Monographs on Statistics & Applied Probability. Boca Raton, FL: Chapman & Hall.

Rue, H., S. Martino, and N. Chopin (2009). Approximate Bayesian inference for latent Gaussian models using integrated nested Laplace approximations (with discussion). *Journal of the Royal Statistical Society: Series B 71*(2), 319–392.

Rue, H., A. I. Riebler, S. H. Sørbye, J. B. Illian, D. P. Simpson, and F. K. Lindgren (2017). Bayesian computing with INLA: A review. *Annual Review of Statistics and Its Application 4*, 395–421.

Rue, H. and H. Tjelmeland (2002). Fitting Gaussian Markov random fields to Gaussian fields. *Scandinavian Journal of Statistics 29*(1), 31–49.

Ruiz-Cárdenas, R., E. T. Krainski, and H. Rue (2012). Direct fitting of dynamic models using integrated nested Laplace approximations — INLA. *Computational Statistics & Data Analysis 56*(6), 1808 – 1828.

Schabenberger, O. and C. A. Gotway (2004). *Statistical Methods for Spatial Data Analysis*. Boca Raton, FL: Chapman & Hall/CRC.

Schlather, M., A. Malinowski, P. J. Menck, M. Oesting, and K. Strokorb (2015). Analysis, simulation and prediction of multivariate random fields with package RandomFields. *Journal of Statistical Software 63*(8), 1–25.

Schmidt, A. M. and A. E. Gelfand (2003). A Bayesian coregionalization approach for multivariate pollutant data. *Journal of Geographysical Research 108*(D24).

Simpson, D. P., J. B. Illian, F. Lindren, S. H. Sørbye, and H. Rue (2016). Going off grid: computationally efficient inference for log-Gaussian Cox processes. *Biometrika 103*(1), 49–70.

Simpson, D. P., H. Rue, A. Riebler, T. G. Martins, and S. H. Sørbye (2017). Penalising model component complexity: A principled, practical approach to constructing priors. *Statistical Science 32*(1), 1–28.

Sørbye, S. and H. Rue (2014). Scaling intrinsic Gaussian Markov random field priors in spatial modelling. *Spatial Statistics 8*, 39–51.

Spiegelhalter, D. J., N. G. Best, B. P. Carlin, and A. Van der Linde (2002). Bayesian measures of model complexity and fit (with discussion). *Journal of the Royal Statistical Society: Series B 64*(4), 583–616.

Stephenson, A. G. (2002). evd: Extreme value distributions. *R News 2*(2), 31–32.

Taylor, B. M., T. M. Davies, R. B. S., and P. J. Diggle (2013). lgcp: An R package for inference with spatial and spatio-temporal log-Gaussian Cox processes. *Journal of Statistical Software 52*(4), 1–40.

Therneau, T. M. (2015). *A Package for Survival Analysis in S.* version 2.38.

Therneau, T. M. and P. M. Grambsch (2000). *Modeling Survival Data: Extending the Cox Model.* New York: Springer.

Tobler, W. R. (1970). A computer movie simulating urban growth in the Detroid region. *Economic Geography 2*(46), 234–240.

Vivar, J. C. and M. A. R. Ferreira (2009). Spatiotemporal models for Gaussian areal data. *Journal of Computational and Graphical Statistics 18*(3), 658–674.

Wang, X., J. J. Faraway, and Y. Yue Ryan (2018). *Bayesian Regression Modeling with INLA.* Boca Raton, FL: Chapman & Hall/CRC.

Watanabe, S. (2013). A widely applicable Bayesian information criterion. *Journal of Machine Learning Research* (14), 867–897.

West, M. and J. Harrison (1997). *Bayesian Forecasting and Dynamic Models.* New York: Springer-Verlag.

Zuur, A. F., E. N. Ieno, and A. A. Saveliev (2017). *Beginner's Guide to Spatial, Temporal and Spatial-Temporal Ecological Data Analysis with R-INLA.* Highland Statistics Ltd.

Index